Heritage of European Mathematics

Previously published
Andrzej Schinzel, Selecta, *Henryk Iwaniec, Władysław Narkiewicz, and Jerzy Urbanowicz* (Eds.)

Thomas Harriot's Doctrine of Triangular Numbers: the 'Magisteria Magna'

Edited by

Janet Beery

Jacqueline Stedall

European Mathematical Society

Editors:

Janet Beery
Department of Mathematics
University of Redlands
Redlands, CA 92373, U.S.A.

E-mail: janet_beery@redlands.edu

Jacqueline Stedall
The Queen's College
University of Oxford
Oxford, OX1 4AW, U.K.

E-mail: jackie.stedall@queens.ox.ac.uk

2000 Mathematics Subject Classification: 01-02

Key words: Harriot, seventeenth century, figurate numbers, interpolation

ISBN 978-3-03719-059-3

The Swiss National Library lists this publication in The Swiss Book, the Swiss national bibliography, and the detailed bibliographic data are available on the Internet at http://www.helveticat.ch.

Contact address:

European Mathematical Society Publishing House
Seminar for Applied Mathematics
ETH-Zentrum FLI C4
CH-8092 Zürich
Switzerland

Phone: +41 (0)44 632 34 36
Email: info@ems-ph.org
Homepage: www.ems-ph.org

Typeset using the author's TEX files: I. Zimmermann, Freiburg
Printed in Germany

9 8 7 6 5 4 3 2 1

Preface

By about 1614, Thomas Harriot (1560–1621) had developed finite difference interpolation methods to aid the construction of navigational tables. In 1618 (or slightly later) he composed a treatise entitled 'De numeris triangularibus et inde de progressionibus arithmeticis, Magisteria magna', in which he derived symbolic interpolation formulae and showed how to use them. Sir Charles Cavendish (1595–1654) still expected in 1651 that the 'Magisteria' would eventually be published: 'till this be printed', he wrote, 'I shall esteem of my owne coppie'. Almost four centuries later, Cavendish's hopes have finally been realized. The pages are presented in this volume in facsimile, and we have added a commentary to help the reader to follow Harriot's beautiful but almost completely nonverbal presentation.

An introductory essay preceding the treatise gives an overview of the contents of the 'Magisteria' and describes its influence on Harriot's contemporaries and successors over the next 60 years. Harriot's method was not superseded until Newton, apparently independently, made a similar discovery in the 1660s. The ideas in the 'Magisteria' were spread primarily through personal communication and unpublished manuscripts, and so, quite apart from their intrinsic mathematical interest, their survival in England during the seventeenth century provides an important case study in the dissemination of mathematics through informal networks of friends and acquaintances.

Contents

Thomas Harriot and the 'Magisteria magna': a short chronology

1560	Harriot born probably in or near Oxford, parentage unknown
1577	registered at St Mary's Hall, Oxford
1584	employed by Walter Ralegh
1585–86	sailed with an English expedition sponsored by Ralegh to the coast of what is now North Carolina; spent a year on and around Roanoke Island
1588	published *A briefe and true report of the new found land of Virginia*
late 1580s	lived on Ralegh's estates in Ireland
1593	came under the patronage of Henry Percy, ninth Earl of Northumberland
1590s	worked on ballistics, optics, physical and alchemical experiments, geometry, algebra, astronomy; none of this published in his lifetime
1607–10	observed [Halley's] comet, satellites of Jupiter, and sunspots
1614	completed tables of meridional parts
1618	wrote out a fair copy of the 'Magisteria magna'
1621	died from cancer of the nose
1651	last known sighting of the original 'Magisteria magna', in the possession of Thomas Aylesbury in Antwerp
1660s	manuscripts thought to be lost
1784	manuscripts rediscovered at Petworth House, Sussex
1810	about 4000 manuscript sheets including the 'Magisteria magna' deposited in the British Museum (later transferred to the British Library); the remaining manuscripts, about 1000 sheets including the tables of meridional parts, remain at Petworth House

Thomas Harriot's 'Magisteria magna' and constant difference interpolation in the seventeenth century

In October 1651, Sir Charles Cavendish, then living in Antwerp, wrote to John Pell in Breda:[1]

> Sr. Th. Alesburie remembers him to you & desires to knowe if you would be pleased to shew the vse of Mr. Hariots doctrine of triangulare numbers; which if you will doe, he will send you the originall; I confess I was so farre in loue with it that I coppied it out; though I doute I vnderstand it not all; much less the many vses which I assure myself you will finde of it.

Thomas Harriot's 'doctrine of triangular numbers' is contained in a manuscript treatise of thirty-eight pages,[2] which was never published but which has survived amongst the several hundred pages of mathematical work left behind by Harriot after his death in 1621. The full title, as Harriot wrote it on the cover sheet, is 'De Numeris Triangularibus et inde De Progressionibus Arithmeticis Magisteria magna T. H.', which may be translated as 'Of triangular numbers and thence of arithmetic progressions, the great doctrine, [by] T[homas] H[arriot]'. The pages that follow contain several tables, numerical and algebraic, and a multitude of equations and formulae, but hardly a word of explanation, and the modern reader is likely to react at first much as Charles Cavendish did in 1651. One has to penetrate only a little way to recognize that there is some very interesting mathematics here, beautifully and concisely written, but it is not immediately easy to see what Harriot was aiming at as he moved apparently so effortlessly from one page to the next, and only in the later stages does the purpose of the treatise become clear. Harriot wrote for himself and a few close friends, who would have already known the context and reason for the work, but the modern reader needs a little more help, and the commentary alongside each page of this edition is designed to elucidate some of the details of Harriot's argument. Before that, this introductory essay offers an overview of the structure and aims of the 'Magisteria' and of its influence on Harriot's contemporaries and successors.

The 'Magisteria' was the document in which Harriot, towards the end of his life, summarized the method he had devised and used some years earlier for interpolating tables by means of constant differences. Its story is of historical importance for at least three reasons. First, Harriot worked out his method in considerable generality, and expressed the relevant formulae in something very close to modern algebraic notation, but it has never been published and the extent of his achievement has not been fully recognized. Second, it is clear from this present study that the methods presented in the 'Magisteria' were the subject of discussion and research amongst English math-

[1] Cavendish to Pell, [26 September]/6 October 1651, British Library Add MS 4278, ff. 321–322; reproduced in Malcolm and Stedall 2005, 584.

[2] BL Add MS 6782, ff. 107–144.

ematicians for a period of about sixty years from 1610 to 1670. As we show later, Torporley, Warner, Pell, Collins, and Mercator, all explored Harriot's methods before or independently of Newton's rediscovery of them in 1665. Apart from a few isolated examples, however, their work on this subject was unpublished and has remained to all intents and purposes invisible to later historians. A topic of enduring interest in early seventeenth-century English mathematical circles has therefore until now passed largely unnoticed.

A third important feature of this story is that it demonstrates how new mathematical ideas were disseminated between small groups of friends, and later passed from one generation to another, without appearing in printed texts. Communication was primarily through verbal explanation and discussion, accompanied by a few key manuscripts that circulated from hand to hand and were borrowed, copied, and talked about, in some cases over periods of up to thirty years. Apart from Harriot and Briggs, the people who wrote, read, or thought about these methods were not outstanding or innovative mathematicians, but all of them worked at the subject seriously and were part of small and fluid mathematical communities that over the years contributed in a variety of ways to the vitality of mathematical studies in England. The personal concerns and relationships of people who gave no thought to recording their activities for the future are never easy to discern, but in this present paper we have a case study that gives us such insights. In studying Harriot's 'Magisteria', therefore, we uncover not only a fascinating and little known piece of seventeenth-century mathematics but also a network of connections and communications between friends and acquaintances, extending in time for over half a century.[3]

The key to Harriot's method, as his title suggests, is an understanding of the properties of what he called 'triangular numbers', and so we begin by looking at what Harriot himself knew or discovered about such numbers.

Triangular numbers

The simplest triangular numbers are those that arise from arranging pebbles or dots in equilateral triangles, and for successively larger triangles are 1, 3, 6, 10, 15, 21, ... If such triangles are imagined stacked vertically in decreasing order of size, they form triangular pyramids (or tetrahedra) and the numbers of pebbles or dots in successive pyramids are clearly sums of consecutive triangular numbers, thus, 1, 4, 10, 20, 35, ... Harriot called these numbers 'pyramidal'. Spatial imagination must end at this point, but arithmetically we may imagine sums of consecutive pyramidals, yielding 1, 5, 15, 35, 70, ..., and so on.[4] Each triangle (or pyramid) is defined by the number of objects

[3]For detailed biographical information on Harriot and his immediate acquaintances see Shirley 1983.

[4]The nomenclature for the successive levels of triangular numbers can be confusing and needs to be understood for each writer individually. Boethius used the description 'pyramis trigona' (triangular pyramid) for a pyramid on a triangular base, constructed from 1, 4, 10, 20, ... units; Maurolico used 'pyramides triangulae primae' for 1, 4, 10, 20, ... and 'pyramides triangulae secundae' for 1, 5, 15, 35, ...; Harriot

in a single side, 1, 2, 3, 4, 5, ..., for which reason these numbers were sometimes called 'laterals' or 'sides' or 'roots'. These in turn are made up from simple units 1, 1, 1, 1, ... The numbers are collectively known as the triangular figurate numbers, or generalized triangular numbers. For brevity we will refer to them here, as Harriot did in his title, simply as 'triangular numbers' (*numeri trianguli*). In Table 1 they are arranged in rows (and columns).

units	1	1	1	1	1	1	1
laterals	1	2	3	4	5	6	7
triangulars	1	3	6	10	15	21	28
pyramidals	1	4	10	20	35	56	84
	1	5	15	35	70	126	210
	1	6	21	56	126	252	462
	1	7	28	84	210	462	924

Table 1. Triangular figurate numbers.

This same array in triangular form, with the 1 from the top left hand corner forming the upper vertex, is now usually known as 'Pascal's triangle' after Blaise Pascal (1623–62), but properties of individual rows and columns and of the array as a whole were known in western Europe from late antiquity onwards.[5] Harriot came across the numbers in a practical context in 1591 when he worked out methods of stacking bullets.[6] Later the triangular numbers entered his mathematical work in several ways, and here we give a brief survey of what he learned about them from existing writings.

The earliest introduction to figurate numbers to circulate widely in western Europe was 'De institutione arithmeticae' of Boethius (*c.* 500 AD). Largely based on an earlier work, the 'Arithmetike isagoge' of Nicomachus (*c.* 100 AD), Boethius' text preserved some elementary Euclidean number theory, and in the later sections introduced figurate numbers: linear, triangular, square, pentagonal, hexagonal, and heptagonal, together with various kinds of pyramidal numbers, and even spherical numbers.[7] Boethius showed how each kind of number can be generated (beautiful coloured diagrams accompany many manuscript copies of his text), and he also gave examples of a few simple relationships, for example, that the sum of two consecutive triangular numbers is a square number. In fact he claimed through this and similar examples that the simple triangular numbers are the basis of all others.

Harriot twice mentioned Boethius as a source of information on triangular numbers,[8] and it is possible that he had access to a manuscript copy of 'De institutione', but he also probably knew the detailed commentary on it published by Jacques Lefevre

described the numbers 1, 4, 10, 20, ... as 'piramidal' and 1, 5, 15, 35, ... as 'triangle piramidall' (BL Add MS 6782, f. 38). Viète, and after him Oughtred and Wallis, described 1, 4, 10, 20, ... as 'pyramidal', 1, 5, 15, 35, ... as 'triangulo-triangular', and 1, 6, 21, 56, ... as 'triangulo-pyramidal'.

[5] For a general history of the triangular figurate numbers, see Edwards 1987.

[6] BL Add MS 6786, ff. 375v–376.

[7] Boethius, Book II, Chapters 5–30.

[8] BL Add MSS 6782, f. 38 and 6784, f. 404v.

d'Etaples (Jacob Faber Stapulensis) in his *Epitome* of 1503. A second edition was printed in Paris in 1522. The *Epitome* offers a detailed comparison of the 'De institutione' of Boethius with the 'De arithmetica' of Jordanus (*c*. 1230). Jordanus was mentioned by Harriot alongside Boethius,[9] so the *Epitome* may well have been his source for both.

A more contemporary treatise, in Boethian style, mentioned by Harriot at least four times,[10] was Francisco Maurolico's *Arithmeticorum libri duo* of 1575. Maurolico's text opens with a section in which he offered several lists of figurate numbers followed by the instructions for generating them. For triangular numbers he gave two methods: adding successive numbers or 'roots' (*per continuatam radicum accumulationem*), or multiplying a 'root' plus one by half of itself (*multiplicando aggregatum collateralis radicis et unitatis in dimidium multitudinis radicum*), in modern notation $(n+1)\frac{n}{2}$. For pyramidals he suggested only the addition of consecutive triangular numbers.[11] In the main part of his text three propositions relate specifically to triangular numbers: Proposition 11, that the sum of two consecutive triangular numbers is a square number; Proposition 54, that eight times a triangular number plus one is a square number; and Proposition 58, that the square of a triangular number is a sum of cubes. All three were demonstrated by Harriot algebraically.[12] Harriot noted that Stevin's commentary on Diophantus and Viète's *Variorum responsorum* also contain the second proposition.[13]

The texts of Boethius and Maurolico go no further than offering lists of numbers and a few elementary relationships. Triangular numbers appear in a more practical context, however, in Michael Stifel's *Arithmetica integra* (1544), with which Harriot was also familiar.[14] Stifel recognized the role of the triangular numbers in the expansion of powers of a binomial, and therefore, conversely, their use in extracting roots (by an extension of the usual algorithm for square roots). Near the beginning of his treatment he offered some 'progressions' of polygonal numbers and displayed some small difference tables, observing, for instance, that the first differences between successive pyramidal numbers are triangular numbers, and that second differences are 'laterals'.[15] Towards the end he gave a table of triangular numbers with seventeen rows.[16]

Girolamo Cardano, in Proposition 137 of his *De proportionibus* of 1570, referred to Stifel's exposition, displayed the same 17-row table, and went on to explain the same methods of finding roots.[17] Fifty pages later, in Proposition 170, the table re-appears, but differently oriented and in a new context: Cardano was now using it to calculate combinations. He noted the additive properties of the table, but also gave a rule for generating the entries by multiplication.[18] For the row 1, 11, 55, 165, 330, 462, 330, 165, …, he gave the following instructions starting from 11:

---

[9] BL Add MS 6782, f. 38.

[10] BL Add MSS 6782, ff. 38, 239, 240; 6787, f. 246.

[11] Maurolico 1575, *a*.

[12] Maurolico 1575, 6, 24, 25; Harriot BL Add MSS 6787, f. 246; 6782, ff. 239–240.

[13] BL Add MS 6787, f. 246. See Stevin 1585, 634 and Viète 1646, 371.

[14] Stifel 1544, 19–46v.

[15] Stifel 1544, 24.

[16] Stifel 1544, 44v.

[17] Cardano 1570, 131–136.

> Or if there are eleven [objects], I want to know quickly the numbers that arise from three choices. First, for the second place, I take 1 from 11 which makes 10, I divide by 2, the number of the place, there comes out 5, I multiply by 11 to make 55, the number of the second place. Next I subtract 2, which is the number of the difference of all the places from the first, from 11, there remains 9, I divide 9 by 3, the number of the place, there comes out 3, I multiply 3 by 55, the second number, to make 165, the number in the third place. Similarly if I want the number of four choices, I take 3, the difference of 4 from the first place, from 11, there is left 8, I divide 8 by 4, the number of the place, there comes out 2, I multiply 2 by 165 to make 330, the number of the fourth place. Similarly for the fifth I subtract 4, the difference from the first place, there remains 7, I divide by 5, the number of the place, there comes out $1\frac{2}{5}$, I multiply by 330, the number in the previous place, to make 462, the number of the fifth place.

The same words (in the original Latin) are copied out in Harriot's manuscripts in a very small neat hand, together with a table of triangular numbers as far as the eleventh row.[19] Thus Harriot knew both from existing writings and from his own investigations that the general triangular numbers appear both in binomial expansions and in calculations of combinations.[20] In his 'Magisteria' he was to find yet another use for them.

Harriot's difference method

From Table 1 it is clear that the lateral numbers exhibit a constant difference (namely, 1), the triangular numbers a constant second difference, pyramidals a constant third difference, and so on. Harriot's theory was concerned with *any* sequence of numbers that has constant first, second, third, or higher differences. (Such sequences arise from polynomial functions, and there is some evidence, to be presented below, that Harriot began to recognize this, but we will start as Harriot did, with numerical sequences only.) A simple example of numbers generated from constant fifth differences, offered by Harriot himself on page 5 of the 'Magisteria', is Table 2, in which each number is the (positive) difference between the two immediately to the right of it.

[18] 'Velut si sint undecim, volo scire breviter numeros qui fiunt ex variatione trium. Primum deduco pro secundo ordine 1 ex 11 fit 10, divido per 2 numerum ordinis, exit 5, duco in 11 fit 55 numerus secundi ordinis. Inde detraho 2, qui est numerus differentiae ordinis totij a primo ex 11, relinquitur 9, divido 9 per 3 numerum ordinis exit 3, duco 3 in 55 numerus secundi fit 165, numerus tertij ordinis. Similiter volo numerum variatione quatuour, deduco 3 differentiam 4 a primo ordine ab 11, relinquitur 8, divido 8 per 4 numerum ordinis, exit 2, duco 2 in 165 fit 330, numerus quarti ordinis. Similiter pro quinto detraho 4 differentiam a primo ordine, relinquitur 7, divido per 5 numerum ordinis exit $1\frac{2}{5}$, duco in 330 numero praecedentis ordinis, fit 462 numerus quinti ordinis.' Cardano 1570, 187. (The pagination of *De proportionibus* is adrift at this point, and runs as follows: 178, 185, 186, 187, 188, 189, 190, 185, 186, 187, 188, 195, 196, ... Proposition 170 as a whole is on the first set of pages numbered 185–187.)

[19] BL Add MS 6782, f. 44.

[20] For Harriot's use of binomial coefficients and expansions for calculating combinations see, for example, BL Add MS 6782, ff. 30, 33–38, 48v, 180–181.

In Table 2, each column (after the first) increases from top to bottom but by subtracting rather than adding the differences the columns can be made to decrease. Harriot tried this too, as in Table 3, which is from page 6 of the 'Magisteria'. He marked increasing columns with the symbol $\triangle$ and/or the letter c for *crescente*, decreasing columns by $\bigtriangledown$ and/or d for *decrescente*, and columns of constant differences with $\square$ and/or e, for *excessus*, meaning the excess, or difference, in an arithmetic progression.[21] Throughout the 'Magisteria' Harriot always made the column starting values (47 and 790, for example, in Table 3) large enough to ensure that within the bounds of his table, he never ran into negative terms, and always took differences in the positive direction so that all the entries appear as positive.[22]

					4
				6	
			7		10
		3		13	
	5		10		23
2		8		23	
	7		18		46
2		15		41	
	9		33		87
2		24		74	
	11		57		161
		35		131	
			92		292
				223	
					515

Table 2. A difference table with increasing columns.

				$\bigtriangledown$
			$\triangle$	
		$\bigtriangledown$		790
	$\triangle$		6	
$\square$		47		784
	5		53	
2		42		731
	7		95	
2		35		636
	9		130	
2		26		506
	11		156	
		15		350
			171	
				179

Table 3. A difference table with increasing and decreasing columns.

[21] The use of these terms is explained by Harriot in BL Add MS 6782, f. 234 and f. 147.

[22] Elsewhere in the manuscripts there are tables containing negative terms, for example, BL Add MS 6782, ff. 165 and 178v.

From two pieces of information, namely (i) the constant difference in the first column and (ii) the first number in each column to the right, it is possible to construct such tables either numerically or algebraically, and Harriot did both. The algebraic version of Table 2 is Table 4, also to be found on Harriot's page 5. (Harriot provided six columns but below, in order to fit it on the page, his table has been truncated after column five.)

				$\triangle$
			$\triangle$	
		$\triangle$		f
	$\triangle$		d	
$\square$		c		$f + d$
	b		$d + c$	
a		$c + b$		$f + 2d + c$
	$b + a$		$d + 2c + b$	
a		$c + 2b + a$		$f + 3d + 3c + b$
	$b + 2a$		$d + 3c + 3b + a$	
a		$c + 3b + 3a$		$f + 4d + 6c + 4b + a$
	$b + 3a$		$d + 4c + 6b + 4a$	
		$c + 4b + 6a$		$f + 5d + 10c + 10b + 5a$
			$d + 5c + 10b + 10a$	
				$f + 6d + 15c + 20b + 15a$

Table 4. An algebraic difference table with increasing columns.

To generate Table 2 from Table 4 we can simply put $a = 2$, $b = 5$, $c = 3$, $d = 7$, $f = 6$, and so on. (Harriot avoided using e here, perhaps because for him, as for Viète, e was reserved for unknown quantities. He did, however, retain a for the constant difference, which has special status in these tables.) Hence, either by repeated addition or by using an extended version of Table 4, we can continue the columns downwards and to the right as far as we like.

Harriot was concerned with a more subtle question: given such a table can we interpolate new values *between* the existing values in a consistent way? The answer is yes, and Harriot produced a set of beautiful algebraic formulae for doing so.

To achieve his purpose, Harriot first needed a clear understanding of the algebraic pattern for each column of Table 4. From inspection it is clear that the numerical coefficients in each column are drawn from Table 5 (sometimes truncated on the right according to the number of terms required).

1						
1	1					
1	2	1				
1	3	3	1			
1	4	6	4	1		
1	5	10	10	5	1	
1	6	15	20	15	6	1

Table 5. Coefficients of the entries in Table 4.

Clearly this is simply a rearranged version of Table 1: the first column consists of units, the second of lateral numbers, the third of triangulars, the fourth of pyramidals, and so on.

The general triangular numbers were therefore fundamental to Harriot's interpolation method. He introduced them right at the beginning of his treatise (pages 1 and 3), and immediately wrote down algebraic formulae for the numbers in the $(n+1)^{\text{st}}$ row of Table 5. Modified slightly into modern notation these were, as we should expect:

$$1, \quad \frac{n}{1}, \quad \frac{n(n-1)}{1.2}, \quad \frac{n(n-1)(n-2)}{1.2.3}, \quad \ldots$$

These, of course, represent precisely the rule given by Cardano in Proposition 170 of *De proportionibus*, but now in general symbolic form. Such algebraic expressions were in themselves a remarkable achievement, the first ever examples of nth-term formulae. They remained unknown outside Harriot's circle, however, until laboriously rediscovered and published in the 1650s by John Wallis.[23]

Harriot's problem may now be described as follows. Suppose we are given a table of numbers as in Table 6, where the third difference is assumed constant, and the first entries in each column are denoted by A, B, C, D (in this context Harriot used capital letters for known quantities, and lower case letters for those he needed to find).

			D
		C	
	B		$D+C$
A		$C+B$	
	$B+A$		$D+2C+B$
A		$C+2B+A$	
	$B+2A$		$D+3C+3B+A$
		$C+3B+3A$	
			$D+4C+6B+4A$

Table 6. An algebraic difference table with known entries.

[23]Wallis 1656, Propositions 171–182. It is very unlikely that Wallis at this date had any knowledge of the contents of Harriot's manuscripts, and his derivations are quite different.

Now suppose we wish to interpolate $n-1$ new values between every entry in the column under D. The first value, D itself, will remain the same, but the first entries in every other column will have to change. Harriot called the new first entries a, b, c, d, and found the following relationships between the original entries A, B, C, D and the new ones a, b, c, d (see page 22 of the 'Magisteria'). For ease of reading, the formulae are given here in modern superscript notation rather than Harriot's nn, nnn, and so on.

$$\begin{aligned} A &= n^3 a \\ B &= n^2 b + (n^3 - n^2)a \\ C &= nc + \frac{n^2-n}{2}b + \frac{n^3-3n^2+2n}{6}a \\ D &= d. \end{aligned}$$

These equations are easily rearranged to give (also on page 22 of the 'Magisteria')

$$\begin{aligned} a &= \frac{A}{n^3} \\ b &= \frac{nB-(n-1)A}{n^3} \\ c &= \frac{6n^2C-(3n^2-3n)B+(2n^2-3n+1)A}{6n^3} \\ d &= D. \end{aligned}$$

Once a, b, c, and d were determined, it was not difficult for Harriot to calculate the Nth entry in the column headed $D(=d)$ (page 26), and he assigned it the label $\frac{N}{n}$ (page 27). This symbol is not to be read as a fraction; it simply denotes the Nth entry of a table interpolated to n times its original length. Thus, for example, if a table is subtabulated to 10 times its original length, the first few new entries will be labelled $\frac{1}{10}, \frac{2}{10}, \frac{3}{10}, \dots$. Entry $\frac{10}{10}$ will be an original entry, and the table will then continue with new entries $\frac{11}{10}, \frac{12}{10}$, and so on. On pages 27 to 33 of the 'Magisteria' Harriot performed a remarkable sequence of algebraic manipulations to provide various versions of the formulae for $\frac{N}{n}$, and ended, on pages 35 and 36, with some worked examples in which they were put to use.

The origins and 'many uses' of Harriot's method

The 'Magisteria', like most of Harriot's work, is undated, but there is a sheet of rough working that gives a clue as to when he wrote out the finished version that now survives. This sheet had previously been intended for other purposes and is headed 'De causa reflexionis ad angulos aequales' ('On the cause of reflection at equal angles'). It is then dated very precisely to half past ten in the morning on Sunday 28 June 1618 (June.28.⨀.ho:10½ ante mer:1618).[24] Harriot must have abandoned this particular

[24]BL Add MS 6784, f. 207. 'Sunday' is indicated by the astronomical symbol for the sun. In 1618, 28 June did indeed fall on a Sunday.

sheet almost immediately, because there is nothing on it relating to the title except (possibly) a small rough circle about 1 centimetre in diameter. The rest of the page has been used for rough working related to the 'Magisteria'. In particular there are several short tables of numbers similar to Table 2 above. The one shown in Table 7, from the bottom right hand corner of the page, contains both numbers and letters (compare this with the 'Magisteria', page 5).

		7		
	2			b
3		$7+2$	a	
	$2+3$			$b+a$
3		$7+4+3$	a	
	$2+6$			
3		$7+6+9$		
	$2+9$			

Table 7. A difference table with both numbers and letters.

In this table the numbers in each column are increasing, but on the same page Harriot experimented with decreasing columns too, creating tables of alternating increasing and decreasing columns of the kind that appear on pages 6 and 7 of the 'Magisteria'.

Eventually Harriot was to take into account all possible combinations of increasing and decreasing columns, and on the left hand side of his page of rough work, he listed the different possibilities, for up to five columns; see Table 8. Tables like this for up to six columns appear in the 'Magisteria' on page 8.

ed
ec
edd
edc
ecd
$\underline{ecc}$
$eddc$
$edcc$
$ecdc$
$\underline{eccc}$
$eddd$
$edcd$
$ecdd$
$\underline{eccd}$
$\dots$
$\dots$

Table 8. Possible arrangements of equal, increasing, and decreasing columns.

There is further work on the reverse of the dated page and on the eight sheets that follow it (in the present ordering of the manuscripts) that can be correlated with the contents of the 'Magisteria', some of it rough, but some of it comprising fairly complete drafts of individual pages.[25] These sheets seem therefore to have been what Harriot used to jot down ideas and formulae as he wrote out his fair copy. This suggests that it was compiled some time after the end of June 1618, after Harriot discarded the sheet on which he was going to write about reflection.

The methods of the 'Magisteria', however, can be traced back to at least 1611, when Harriot's friend William Lower wrote to him about his 'doctrine of differences of differences or triangular nombres'.[26] Amongst Harriot's manuscripts there are several sheets headed 'Differentiae differentiarum' ('Differences of differences') or 'Differentiae differentiarum differentiarum', and so on.[27] Some are paginated, as though intended to be read in a certain order, but they are interspersed amongst rough work on the same theme. Many of the sheets, whether well-written or rough, contain algebraic difference tables with numbers substituted into them, as though Harriot was experimenting, and checking his findings arithmetically. These may therefore represent some of his earliest investigations. In most of the tables the notation is a little different from that in the 'Magisteria', the most common convention being $b, c, d, \ldots$ for the original table entries and $a, e, o, \ldots$ for interpolated values. It is likely that Harriot tried different notations at different times and used whichever he liked best in a given context. Unfortunately none of the pages is dated, but if they indeed contain some of the work referred to by Lower, they must have been written in or before 1611.

Harriot wrote out an example of constant difference interpolation for Lower, possibly in response to his letter of 1611, but certainly before Lower's death in 1615.[28] The cover sheet is headed 'Of unaequall progression of sines. For S.W.L.'. The first table (Table 9) is not actually a table of sines at all but to exemplify the method it is written to look like one. Beside the table Harriot wrote 'This example of sines I have so set down as though the numbers were answerable to arkes [angles]. That by it you may see the use of the probleme for sines.'[29] In tables of sines (or cosines) the absolute values of the differences increase and decrease in alternate columns.[30] Harriot, however,

[25]BL Add MS 6784, ff. 207–215. There is further similar work, but undated, in BL Add MS 6782, ff. 193–220.

[26]Lower to Harriot, 1611, BL Add MS 6789, f. 429.

[27]BL Add MSS 6782, ff. 244v; 6783, f. 45v; and in 6787, ff. 338v–349v.

[28]BL Add MS 6787, ff. 245–252. The main working appears on ff. 247, 248, 250, 252 with rough work on ff. 249 and 251. Folio 246 contains two propositions on triangular numbers from Maurolico; it is in much smaller handwriting than the surrounding sheets and appears to have been inserted into this batch of notes from elsewhere.

[29]For a further example of a similar kind, see also BL Add MS 6787, f. 58, where there is a table of 'pretend' sines into which Harriot interpolated two new values. Folios 59–63 have difference tables for real sines, but in the fourth column the differences become ragged, even when the sines themselves are taken to nine figures. This may be why Harriot chose to demonstrate the method to Lower with 'pretend' sines. Harriot did show how to interpolate a table of real sines at f. 59 (with calculations at f. 61) using the technique demonstrated on f. 58. Both ff. 59 and 61 are headed 'Ad calculum sinuum, per progressiones'.

[30]The absolute values of successive derivatives of $\sin x$ are $|\cos x|, |\sin x|, |\cos x|, \ldots$, which alternately decrease and increase as x increases from 0 to $\frac{\pi}{2}$.

deliberately constructed this table with constant second differences. He marked with an asterisk ⋆ a row in which he was particularly interested.[31]

	△	▽	□	
10″	90			
		1245		
20″	1335		300	
		945		
30″	2280		300	⋆
		645		
40″	2925		300	
		345		
50″	3270			

Table 9. A table of 'pretend' sines.

Harriot then demonstrated the interpolation of nine new values (or ten new spaces) between 30″ and 40″. The rules were essentially the same as those later presented in the 'Magisteria', though the notation here is a little different. Harriot always used n for the number of interpolated spaces, but here he used e for the new second difference (rather than a), and p for the new leading first difference (rather than b).[32]

First, putting $n = 10$ in the formula $enn = 300$, Harriot calculated

$$e = \frac{300}{100} = 3.$$

His formula for the first difference now gave him:[33]

$$pn + \frac{en - enn}{2} = 645,$$

from which he derived $p = 78$. The new interpolated table starting from 30″ is therefore as shown in Table 10.[34]

---

31 BL Add MS 6787, ff. 247 and 248.

32 On a page immediately following the 'Magisteria' in the present ordering of the manuscripts, BL Add MS 6782, f. 145, Harriot sketchily correlated the notation of the 'Magisteria' with the e, p, $\overset{2}{p}$, $\overset{3}{p}$ notation found elsewhere in his writings. There is a draft of this page also at BL Add MS 6784, f. 171.

33 This formula appears on pages 18 and 22 of the 'Magisteria', but there are adjustments to the signs here because the first differences are decreasing rather than increasing.

34 BL Add MS 6787, f. 248.

		△		
★	30″	2280	▽	
			78	□
	31″	2358		3
			75	
	32″	2433		3
			72	
	33″	2505		3
			69	
	34″	2574		3
			66	
	35″	2640		3
			63	
	36″	2703		3
			60	
	37″	2763		3
			57	
	38″	2820		3
			54	
	39″	2874		3
			51	
★	40″	2925		3

Table 10. 'Pretend' sines from Table 9, now interpolated with nine new entries.

The interpolation of trigonometric tables was clearly one possible use of Harriot's difference method. Harriot denoted sines by the symbol υ, secants by ψ (to represent one line cutting another), and tangents by a small circle touching a line, and these symbols appear repeatedly in the 'Magisteria' next to the relevant patterns of increasing and decreasing columns. Actual examples of such interpolation, however, are rare.[35]

Harriot also worked for many years on the calculation of meridional parts (the adjustments needed at each degree of latitude in order to calculate an accurate position on a constant compass bearing).[36] His tables of meridional parts and also some of the preliminary tables show many difference calculations, sometimes as far as the fifth difference. A calculation very similar to the one Harriot wrote out for Lower appears in connection with these preliminary calculations, on five explanatory pages,[37] labelled by Harriot as *ac*.1 to *ac*.5. Harriot began on *ac*.1 with Table 11, where the

[35] Difference tables of sines are to be found in, for instance, Petworth MS 241.5, f. 12, and BL Add MSS 6782, f. 194, and 6787, ff. 59–63, and difference tables of secants at Petworth MS 240, ff. 318, 320, 321, and BL Add MS 6787, f. 64. Harriot developed formulae for interpolating tables of sines ('Ad calculum sinuum') in Petworth MS 241.5, ff. 9–10, and BL Add MS 6787, ff. 17–20, 53–57, 66–71, 252. Sine or 'pretend' sine interpolation occurs in BL Add MS 6787, ff. 58–59, 61, 72–73, 247–248, 250, and Petworth MS 241.5, ff. 11–15; and 'pretend' secant interpolation in BL Add MS 6787, f. 63v.

[36] For details of these calculations see Pepper 1968.

entries for each minute are successive integer powers of what we would now write as 0.999 709 154 09.... The first differences, all beginning 2,90 ... (= 0.000 290 ...), are written in the central column and repeated on the right, and the second difference is taken to be constant at 8460 (= 0.000 000 084 60). Harriot took his unit to be an appropriate power of ten, so that his entries appear as integers, with groups of digits separated by commas for ease of reading; these have been retained in Table 11 as he wrote them.

	10, 002, 909, 305, 1	
	2, 909, 305, 1	2, 909, 305, 1
$0'$	10, 000, 000, 000, 0	8460
	2, 908, 459, 1	2, 908, 459, 1
$01'$	9, 997, 091, 540, 9	8460
	2, 907, 613, 1	2, 907, 613, 1
$02'$	9, 994, 183, 927, 8	

Table 11. Powers of 0.999 709 154 09 (9,997,091,540,9) tabulated for every minute.

Harriot then proposed to interpolate 99 new values between $00'$ and $01'$ ('sit progressio dividenda in 100 partes', 'let the progression be divided into 100 parts').[38] As in his calculation for Lower, he denoted the first and second differences in the new table by p and e, respectively, and from the formula $enn = 8460$ he calculated

$$e = \frac{8460}{10000} = \frac{846}{1000}.$$

Next, from

$$\frac{2pn + en - enn}{2} = 2,908,459,1$$

he calculated

$$p = 29,088,7,787.$$

The new interpolated table must therefore begin as shown in Table 12.

From now on Harriot could construct the rest of the table simply by adding differences, but he could also calculate values using his formulae, and on sheets *ac*.2 to *ac*.5 he showed how to do this for the 10th, 20th, 30th, ..., 100th entries.[39]

[37]Petworth MS 240, ff. 277–281. As often in Harriot's manuscripts, the present ordering is Harriot's in reverse, thus *ac*.5 to *ac*.1. These five sheets follow, and are related to, what Pepper has called Harriot's γ-tables, Petworth MS 240, ff. 272–276. Pepper has denoted by β (= 0.999 709 154 09) what he calls the 'fundamental constant' for Harriot's logarithmic spiral. The 'fundamental table' tabulated for each minute consists of integer powers of β, but when Harriot wanted to tabulate for each hundredth of a minute he needed to calculate the quantity Pepper has called γ, namely the $\mathbf{100}^{\text{th}}$ root of β. This he found by interpolation to be = 0.999 997 091 122 13. For details see Pepper 1968, 375–379 and 405–406.

[38]The full calculation is reproduced in Pepper 1968, 405–406, but Pepper has read 'dividenda' as 'dimindienda'.

[39]Part of *ac*.3 is reproduced in Pepper 1968, 405.

0′00	10, 000, 000, 000, 0, 000	
	29, 088, 7, 787	29, 088, 7, 787
01	9, 999, 970, 911, 2, 213	0846
	29, 088, 6, 941	29, 088, 6, 941
02	9, 999, 941, 822, 5, 272	0846
...		

Table 12. Powers of 0.999 709 154 09 tabulated for every hundredth of a minute.

Pepper has suggested that Harriot's tables were completed in 1614,[40] in which case, as the example for Lower also suggests, his difference method was worked out well before that date. There are many other examples of n, e, p formulae scattered through the manuscripts, but none is dated.[41] We can safely say, however, that Harriot began thinking about constant difference interpolation by 1611 at the latest, and that by 1614 he had derived algebraic formulae that allowed him to apply the method very efficiently to the interpolation of numerical tables.

Harriot also recognized a number of other potential uses of his method. He used it, for example, to generate a variety of polyhedral numbers, both numerically and algebraically. This work appears in the manuscripts directly after the 'Magisteria'.[42]

There are also two interesting pages which suggest that Harriot noticed an application of a quite different kind. The first is a page on which he experimented with evaluating polynomial expressions for the first few integers.[43] First he calculated values of 'a cube plus its root' or, as he wrote it, $1C + 1R$, for $R = 1, 2, 3, 4, 5$. This is followed by similar tables for $1C + 2R$ and $1C + 3R$. For all three tables Harriot also calculated first, second, and third differences. His table for $1C + 3R$ is Table 13.

	$1C + 3R$			
		4	−6	6
1.	$1 + 3 = 4$	4	0	6
2.	$8 + 6 = 14$	10	6	6
3.	$27 + 9 = 36$	22	12	6
4.	$64 + 12 = 76$	40	18	6
5.	$125 + 15 = 140$	64	24	6

Table 13. A difference table for a cube plus three times its root, for $R = 1, 2, 3, 4, 5$.

[40] See Pepper 1967, 39, or Pepper 1968, 359, 365–366, 385.

[41] Formulae in n, e, p notation, and associated working, have been identified in the following runs of manuscripts (though the list may not be exhaustive): BL Add MSS 6782, ff. 145, 147–148, 165–178v, 198, 234–236; 6784, f. 171; 6785, f. 84v; 6787, ff. 17–20, 53–58, 66–73, 249–252; 6789, ff. 102v–103; Petworth MSS 240, ff. 277–281; 241.5, f. 11. Janet Beery has investigated the possible ordering of this and related material with a view to reconstructing Harriot's method of discovery; see Beery 2007.

[42] BL Add MS 6782, ff. 147–159.

[43] BL Add MS 6782, f. 246.

The first, second, and third differences are shown in the three columns on the right. Those below the stepped line are found directly from the calculated values of 4, 14, 36, 76, 140, but those above the steps appear to have been obtained by extrapolating upwards, using the fact that the third difference is always 6. To the right of this table, Harriot constructed a second one (Table 14), this time including $R = -2, -1, 0$, as though to check that the pattern he had just found by extrapolation was correct.

$-2.$	$-8 - 6 = -14$	22	-18	6
$-1.$	$-1 + -3 = -4$	10	-12	6
0.	$0 + 0 = 0$	4	-6	6
1.	$1 + 3 = 4$	4	0	6
2.	$8 + 6 = 14$	10	6	6
3.	$27 + 9 = 36$	22	12	6

Table 14. A difference table for a cube plus three times its root, for $R = -2, -1, 0, 1, 2, 3$.

Clearly, Harriot could extend the table upwards or downwards as far as he wished, either by simply adding differences, or by using his formulae. The relevant formulae are written out in n, e, p notation at the bottom of the page, below the numerical tables, and are the same as those he used to calculate 10th, 20th, 30th, ..., entries in Table 12. Thus Harriot had a method of solving the equation $1C + 3R = N$ for a given number N, provided the root was an integer. For non-integer roots he would have needed to interpolate, but his formulae enabled him to do that too.

The second page on which such material appears gives polynomial expressions which now contain cubes (C), squares (Z), and roots (R), suggesting that Harriot was indeed concerned here with equation solving.[44] The first of the tables from this second page is Table 15.

	$1C + 1Z + 1R$			
1.	$1 + 1 + 1 = 3$	3	2	6
2.	$8 + 4 + 2 = 14$	11	8	6
3.	$27 + 9 + 3 = 39$	25	14	6
4.	$64 + 16 + 4 = 84$	45	20	6
5.	$125 + 25 + 5 = 155$	71	26	6

Table 15. A difference table for a cube plus a square plus a root.

There are three further similar tables on the same page. The four polynomial expressions and the numbers from the first rows of their respective tables are shown in Table 16.

[44] BL Add MS 6782, f. 254.

$1C + 1Z + 1R$	3	3	2	6
$1C + 2Z + 3R$	6	6	4	6
$1C + 3Z + 2R$	6	6	6	6
$1C + 4Z + 5R$	10	10	8	6

Table 16. Polynomials, with the first rows of their difference tables.

Denoting a first row by $a\ b\ c\ d$, it is very easy to see by inspection of Table 16 that $a = b =$ the sum of the coefficients, $c =$ twice the coefficient of Z, and $d = 6$. A few more examples would support these rules strongly enough for any seventeenth-century mathematician to believe them (and for us a quick check by algebra proves them). Further, from a first row, and the constant difference 6, an entire table can be generated without ever needing to evaluate the polynomial (except trivially for $R = 1$, by adding its coefficients). Harriot did not write down rules for the first row, but those who came after him certainly did, as we will see later, and it is very likely that Harriot too had seen them.

Another of Harriot's uses of the difference method appears also to have been related to his experimentation with polynomials, and is found in nearby pages in the manuscripts. This is his discovery of formulae for sums of squares, or cubes, or higher powers.[45] Consider, for example, Table 17.

$\overset{4}{p}$	$\overset{3}{p}$	$\overset{2}{p}$	p		
1	1	1	0	6	1
9	8	7	6	6	2
36	27	19	12	6	3
100	64	37	18	6	4
225	125	61	24	6	5

Table 17. A table for finding sums of cubes.

The figures under $\overset{3}{p}$ are cubes. The columns to the right contain successive differences, as far as the constant difference 6. The leftmost column, headed $\overset{4}{p}$, continues the same pattern to the left, and therefore contains sums of cubes: 1, 9, 36, ... This was what Harriot was interested in, because above the table he has written 'Ad summam C' ('For summing cubes'). Since the entire table can be read as a difference table with a constant fourth difference, Harriot knew the formula for each entry under $\overset{4}{p}$. By substituting the appropriate values into the formula, and writing s for the sum of the first a cubes, he could therefore calculate that

$$s.24 = 6, aaaa + 12, aaa + 6, aa + 0, a$$

or

$$s.4 = 1, aaaa + 2, aaa + 1, aa + 0, a.$$

[45] BL Add MS 6782, ff. 239–240.

As a check on this formula Harriot noted a rule given by Maurolico which states that a sum of cubes is the square of the corresponding triangular number.[46] Harriot used the same method to derive formulae for sums of consecutive numbers, squares, and fourth powers, and clearly he could have extended it to higher powers if he had wanted to. Such formulae became crucial to the development of integration theory when they were rediscovered by Fermat and others a few years later.[47]

The influence of the 'Magisteria'

Harriot's interpolation method, though never published, was taken up by some of his closest friends and was to remain in use amongst English mathematicians until the 1660s. In this section, therefore, we trace its survival and use during the fifty years after his death.

Nathaniel Torporley (1564–1632)

Nathaniel Torporley, who shared and understood Harriot's mathematical interests better than anyone, was educated at Oxford slightly after Harriot, from 1581. The date of their meeting and the subsequent course of their friendship is unknown, but a letter that Torporley wrote to Harriot on the eve of his first meeting with Viète, in or shortly after 1600, indicates that both he and Harriot were already familiar with Viète's mathematics.[48] Torporley then became an assistant to Viète, until the latter died in 1603, and in 1608 he became vicar of Salwarpe, in Shropshire. Thereafter he was probably in less regular contact with Harriot. Nevertheless, Harriot chose Torporley in his will as the editor of his mathematical papers. Torporley resigned from his position at Salwarpe in 1622, presumably to carry out the task assigned to him. He spent the final years of his life at Sion College, London, where he died in 1632, and his surviving papers were deposited in the library there. Some were destroyed by fire in 1666; those that remained were transferred to the library of Lambeth Palace in 1996.

A further important document in Torporley's hand, however, survived separately, and in the early eighteenth century was placed by William Jones in the library of the Earl of Macclesfield. It is entitled 'Na. To. CONGESTOR ... eodem se forte resolvit CONIECTOR ...', and is immediately followed by, and bound with, 'Coniector Liber II'. This identifies it as a treatise listed in the Sion College Benefactors' Book as 'Congestor analiticus cui accessit conjector', which no longer survives amongst the Sion College manuscripts. It is *not* the same as another document that did remain at Sion College and which has become known as the 'Congestor', but which more likely corresponds to the item listed in the Benefactors' Book as 'Problemata varia de analiticis'.[49] Confusion has arisen because until recently the Macclesfield papers were

[46]BL Add MS 6782, f. 240; Maurolico 1575, 25.

[47]See Mahoney 1994, 229–233.

[48]BL Add MS 6788, f. 117. The date of the letter is suggested by its reference to Viète as 'Apollonius Gallus'. Viète's book of that name was published in Paris in 1600.

not accessible to scholars. However, they were purchased by Cambridge University Library in 2000 and made publicly available about three years later, so that Torporley's treatise can now for the first time in many years be seen and studied.[50] We do not know whether the 'Congestor … [et] … coniector' from the Macclesfield library is the copy originally held at Sion College, but it seems likely that it is, and that it was 'borrowed' from there in the seventeenth century. The treatise is altogether 164 pages long in densely written text, with intricate tables and diagrams on almost every page, and it seems very unlikely that Torporley in old age would have made any other copy. It is dedicated to the Earl of Northumberland, and dated 5 October 1627.[51]

The spine of the blue board cover that has been added to this as to many items in the Macclesfield collection carries the title 'Of differences', but in a different and later hand than Torporley's, and we shall use Torporley's own title of 'Congestor … [et] … coniector'. Here we are concerned only with the first and longer part of the treatise, the 'Congestor', which runs to 98 pages, and is an explanation of figurate numbers and their use in the interpolation of tables, precisely the subject of Harriot's 'Magisteria'. Torporley mentioned Harriot explicitly in several places,[52] and the similarity to Harriot's work is clear from the start.

On page 4 Torporley gave a table of generalized triangular numbers, similar to Table 4 above, but running to 6 columns and 10 rows (later, on pages 28 to 33, he extended the table to 8 columns and 305 rows). Then on page 5 he presented a numerical table, of which only the first few rows are given in Table 18; Torporley's version extends for a further sixteen lines:

						11
					11	
				22		11
			38		22	
		51		44		11
1.	85		82		33	
		133		77		11
2.	218		159		44	
		292		121		11
3.	510		280		55	
		572		176		11
4.	1082		456		66	
		1028		242		11
5.	2110		698		77	
		1726		319		11

Table 18. Part of Torporley's difference table, with increasing columns.

[49] 'Problemata varia de analiticis', sometimes known as 'Congestor': Sion College MS Arc L.40.2/L.40, ff. 1–34v. The misidentification of this document stems from Tanner 1977, and it has only been possible to clarify the matter since the Macclesfield papers became publicly available in 2003.

[50] 'Congestor … [et] … coniector': now CUL Add MS 9597/17/28.

[51] 'Ex bibliotheca tua Petworthia. V Kal: Octob: 1627', CUL Add MS 9597/17/28, iii verso.

[52] Pages iii (twice), viii, 2, 69 (twice), 98 (four times).

Each column is increasing, and Torporley noted that this is what happens in tables of tangents and secants, and also of antilogarithms ('Antilogarithmi Nepieri'). On the next page he gave an example of a similar table in which the columns alternately increase and decrease, and noted that this is the pattern for tables of sines. On page 6, in further discussion of possible patterns he also mentioned two patterns for logarithms, one for those originally defined by Napier, which decrease from $\log 0 = \infty$ to $\log 10^7 = 0$, and another for those of Briggs, which increase from $\log 1 = 0$ to $\log 10 = 1$.

On page 44, now almost halfway through his treatise, Torporley produced a constant difference table in which the entries are written algebraically, see Table 19. His notation here was different from Harriot's (see Table 4 above) in that he denoted the constant difference by c and the leading entries in the columns leftwards by b and a respectively. To take account of both added and subtracted differences he used a dot to represent either plus or minus, so that $1a.1b$ is to be read as $a \pm b$.

$0a.0b.0c$			
	$1.a$		
$1.a$		$1.b$	
	$1a.1b$		$1.c$
$2a.1b$		$1b.1c$	
	$1a.2b.1c$		$1.c$
$3a.3b.1c$		$1b.2c$	
	$1a.3b.3c$		$1.c$
$4a.6b.4c$		$1b.3c$	
	$1a.4b.6c$		$1.c$
$5a.10b.10c$		$1b.4c$	
	$1a.5b.10c$		$1.c$
$6a.15b.20c$		$1b.5c$	
	$1a.6b.15c$		$1.c$
$7a.21b.35c$		$1b.6c$	
	$1a.7b.21c$		$1.c$
$8a.28b.56c$		$1b.7c$	
	$1a.8b.28c$		
$9a.36b.84c$			

Table 19. Torporley's algebraic difference table.

Torporley followed this with numerical examples of such patterns, one taken from a table of sines (page 48) and one from a table of logarithms (page 56). His example of sines is Table 20 (where 'g' stands for *gradus*, or degree).

g	′	″	sinus	Diff pr[ima]	Diff 2nd
0	38	1	1105 83746 69929		
				4847840238	
0	38	2	1106 32225 10167		260
				4847839978	
0	38	3	1106 80703 50145		260
				4847839718	
0	38	4	1107 29181 89863		

Table 20. A difference table for sines.

On page 69 Torporley made some remarks on notation and introduced Harriot's inequality signs. He also observed that we may write a^{VI} in place of $aaaaaa$, and $b^{IV}a^{IV}$ in place of $bbbbaaaa$, the earliest known use of superscript notation for powers.[53] From this point onwards Torporley moved away from the kind of numerical examples he had given until now, and began to give general algebraic formulae of the kind Harriot produced in the 'Magisteria'. Unfortunately, despite just having introduced Harriot's notation, Torporley reverted to different notation of his own, which was not nearly so clear. Where Harriot used n Torporley used sometimes m or x, but most often a squiggly figure which is most nearly reproduced by the Greek letter ξ. When this was squared or cubed he preceded it by $\hat{2}$ or $\hat{3}$; thus $\hat{2}\xi$ may be written in modern notation as n^2. Although Torporley retained Harriot's use of capital letters A, B, C, … for the first entries of each column of the original table and a, b, c,… for the corresponding entries in the interpolated tables, he reversed Harriot's ordering, so that for Torporley A is the first entry of the column to be interpolated, while B, C, D, … head the columns of second, third, fourth differences, respectively.

Bearing in mind all these changes of notation, it can be seen that Torporley nevertheless reproduced Harriot's formulae. On page 73, for instance, he wrote a table that begins as shown in Table 21.

If we reverse the order of A and B in the first row, and A, B, and C in the second, then we have the formulae given by Harriot on page 25 of the 'Magisteria' under the headings 'Canones *c*' (*edc* and *edd* pattern) and 'Canones *d*' (*ecdc* pattern), respectively. Torporley used such formulae in all that followed, up to the end of his treatise some thirty pages later.

On the final page Torporley acknowledged his debt to Harriot, and wrote that Harriot gave three different titles to his work on this subject. For the first two Torporley quoted the title page and endpage of the 'Magisteria', but the third, as it appeared 'abundantly in lost sheets' (*Sed in chartula seorsim perdita copiosus*), was 'Poristicum Arithmeticarum Progressionum', which may be (roughly) translated as 'The finding out of arithmetic

[53] Torporley also used this notation when copying Harriot's equations. There are many examples of it in Sion College MS Arc L.40.2/L.40, ff. 42–54v. Bombelli and Stevin had both devised a 'number-in-a-circle' notation for powers of a single, nameless, unknown; Viète introduced letters A, E, … for an unknown but had no general notation for powers. Torporley's (and modern) superscript notation labels the unknown *and* its power.

progressions'. Harriot, according to Torporley, claimed that such progressions could be used for tables of sines, and of figurate numbers, to find arcs and sines to any degree of accuracy, and sections of angles. None of which, said Torporley, had been satisfactorily explained previously, except figurate numbers, which most writers since Boethius had regarded as useless.

$+a$	$-b$	$+c$
$2\xi A - B$	$\underline{2B}$	
$\underline{\xi B}$	$2\hat{2}\xi$	
$2\hat{2}\xi$		
$6\hat{2}\xi A - 3\xi B$	$6\xi B - 6\xi C$	$\underline{6C}$
$3\hat{2}\xi B - 2\hat{2}\xi C$	$\underline{6C}$	$6\hat{3}\xi$
$\underline{3\xi C - C}$	$6\hat{3}\xi$	
$6\hat{3}\xi$		

Table 21. Torporley's interpolation formulae.

It is hard to imagine two treatises more different in style than Harriot's 'Magisteria', concise and almost entirely symbolic, and Torporley's 'Congestor', rambling, wordy, and long-winded. Nevertheless it is clear that Torporley was treating essentially the same subject matter. He aimed at, but never achieved, the algebraic generality that Harriot mastered so brilliantly, but on the other hand he did explain the purpose of the method and showed how to make the necessary calculations.

Torporley must have begun the 'Congestor' very soon after he started work on Harriot's papers because he sent material almost identical to some now found on his pages 37 and 38 to John Protheroe, one of Harriot's mathematical executors, before 1624.[54] By 1627 the treatise was complete and if it was, as we suppose, the document recorded in the Benefactors' Book, it must have accompanied Torporley to Sion College. What happened to it after that is something of a mystery, for there is no further mention of it until it reappears in the hands of John Collins over forty years later. It is likely that on or after completion it was shown to Aylesbury, the sole surviving mathematical executor, and perhaps Aylesbury later borrowed it from Sion College on behalf of Warner, who was developing his own project using the method of differences (see below). This might explain how it came into the hands of Collins, for after Warner died his mathematical papers eventually passed to Herbert Thorndike, who in 1667 lent some of them to Collins.[55] Unfortunately, the inventory of Warner's papers made out by Collins on 14 December 1667 makes no mention of the 'Congestor' or anything that can be identified with it, but it cannot be ruled out that Thorndike gave it to him

[54]The pages are now to be found amongst Warner's papers in BL Add MS 4395, ff. 89–90. Presumably either Aylesbury or Warner acquired them from Protheroe's widow after Protheroe's death in 1624.

[55]BL Add MS 4394, f. 106, and printed in Halliwell-Phillips 1841, 95.

on some other occasion. We know that Collins had it in his possession by the end of 1668, because in January 1669 John Wallis wrote to Collins as follows:[56]

> … he must, at convenient distances, make essays whether the first, second, third, fourth, or fifth differences will serve the turn; and then it will be easy to apply them in such manner as Torporley, if I remember the name aright, in the manuscript you shewed me.

Collins mentioned the document again in a letter written probably to Gilbert Clark, and probably in the early 1670s:[57]

> I can tell you, by the perusal of some papers of Torporley's it appears that Harriot could make the sine of any arch at demand, and the converse, and apply a table of sines to solve all equations, and treated largely of figurate arithmetic.

Collins' statement seems to refer to the paragraph on the final page of the 'Congestor', except that there is no mention there of solving equations. Collins was certainly in no doubt that the contents of Torporley's treatise were to be attributed to Harriot.

In the early years of the eighteenth century, Collins' books and papers, including his copy of the 'Congestor … [et] … Coniector', were acquired by William Jones, tutor to the son of the Earl of Macclesfield. In 1713 Jones lent some papers to Roger Cotes, with the following accompanying letter[58]

> I waited for an opportunity of sending you some old manuscripts I had by me, and at last am obliged to venture them by the Carrier; they relate, in some measure, to the Method of Differences: the folio one, I find, was writ by one Nath. Torporley, a Shropshire man, who when young was Amanuensis to Vieta, but afterwards writ against him; he was contemporary with Briggs and Harriot, and intimately acquainted with them: the Book, I think, can be of no other use to you, than in what relates to the History of that Method, and in having the satisfaction of seeing what had bin formerly done on that subject. The other small 4to is a piece of Mercator's about Differences.

We shall return to the small quarto piece by Mercator later, but the folio manuscript by Torporley was undoubtedly the 'Congestor'. Cotes later returned it to Jones, and it remained with Jones' other books and papers in the library of the Earls of Macclesfield until acquired by Cambridge University Library in 2000.

In 1713 Jones remarked that the 'Congestor … [et] … Coniector' was by now of value only as a historical document. Was it ever more than that? There is no evidence that it was read by anyone for many years after it was written, except perhaps by Aylesbury or Warner, but since both of them had direct access to Harriot's original papers they hardly needed to learn from Torporley. By 1668 it had somehow passed to John Collins, who in turn showed it to Wallis. As we have seen, Wallis commented in January 1669 that it might be useful to refer to it on methods of differences. It is unlikely that anyone took up his suggestion, since mathematicians like Pell and Mercator who were interested

[56]Rigaud 1841, II, 511.

[57]Rigaud 1841, II, 478.

[58]Newton Collection, Trinity College, Cambridge, R.16.38 (b), 315; quoted in Tanner 1977, 403.

in such methods had by now worked out their own ways of proceeding (see below). Indeed, it is hard to imagine that anyone could ever have tried to learn from Torporley's long-winded and convoluted prose, and his 'Congestor' seems to have had no influence. In that sense, our investigation of its contents brings us to a dead end.

Henry Briggs (1561–1630)

There is just one further trail to follow in assessing Torporley's influence on his contemporaries. We have already seen that on page 6 of the 'Congestor' Torporley mentioned the logarithms of both Napier and Briggs. His own example of logarithms on page 56 is preceded by a section of text addressed directly to Henry Briggs, whom Torporley called 'my friend and noted mathematician' ('amicum meum ac Mathematicum insignem'); he also noted that Briggs gave him a copy of his *Logarithmorum chilias prima*, published in 1617. From 1615 onwards, Briggs was as deeply involved in the construction of tables of logarithms as Harriot had earlier been in the construction of tables of meridional parts, and if there was one mathematician in England who might be expected to understand the problems of interpolation at least as well as Harriot, it was Briggs.

In his *Arithmetica logarithmica*, published in 1624, Briggs certainly suggested a constant difference method as a possible way of interpolating new entries into a table of logarithms.[59] He did not explain any theory, but illustrated the procedure with a worked example as follows: given the logarithms of 91235 and 91236, find the logarithms of the nine intermediate numbers that we may write in modern decimal notation as 91235.1, 91235.2, ..., 91235.9. This, of course, was precisely the problem Harriot had been concerned with in the 'Magisteria'. Briggs' answers are identical to those that would be obtained using Harriot's formulae, but his method is a little different, at first sight less theoretical and rather more intuitive. Briggs' initial table is shown in Table 22. There the letters A, B, C, D indicate numbers (A), and their logarithms (B), first differences (C), and second differences (D). (Although Briggs claimed to be starting only from logarithms of 91235 and 91236 he must also have used logarithms of 91234 and 91237, to calculate the three first differences marked C.)

	47602, 0016.C	
91235.A.	4, 96016, 14763, 8639.B	5217.D
	47601, 4799.C	
91236.A.	4, 96016, 62365, 3438.B	5217.D
	47600, 9582.C	

Table 22. An extract from a table of logarithms, to be interpolated.

Briggs now instructed that 5217 (D) should be multiplied by 5, 15, 25, 35, 45 to produce numbers he called K, I, H, G, F, as shown in Table 23.

[59] Briggs 1624, 24–26.

F	234 765	45
G	182 595	35
H	130 425	25
I	78 255	15
K	26 085	5

Table 23. Briggs' quantities K, I, H, G, F.

Briggs divided these quantities by 100 (by inserting two extra places in front of the leading figure) before adding or subtracting them to or from the central first difference C and dividing by 10 (see Table 24). Thus he effectively added or subtracted $\frac{5}{1000}$, $\frac{15}{1000}$, $\frac{25}{1000}$, $\frac{35}{1000}$, $\frac{45}{1000}$ of D to or from $\frac{1}{10}$ of C, so that the new second difference is $\frac{1}{100}$ of D, just as Harriot's theory had demonstrated that it should be.

Number	Logarithm	
912350	4, 96016, 14763, 8639	
	4760, 1715	$C + F$
912351	4, 96016, 19524, 0354	
	4760, 1662	$C + G$
912352	4, 96016, 24284, 2016	
	4760, 1610	$C + H$
912353	4, 96016, 29044, 3626	
	4760, 1558	$C + I$
912354	4, 96016, 33804, 5184	
	4760, 1506	$C + K$
912355	4, 96016, 38564, 6690	
	4760, 1454	$C - K$
912356	4, 96016, 43324, 8144	
	4760, 1402	$C - I$
912357	4, 96016, 48084, 9546	
	4760, 1350	$C - H$
912358	4, 96016, 52845, 0896	
	4760, 1297	$C - G$
912359	4, 96016, 57605, 2193	
	4760, 1245	$C - F$
912360	4, 96016, 62365, 3438	

Table 24. The logarithms from Table 22, now interpolated with nine new entries.

Briggs did not offer any justification for this series of steps. His method was altogether more pragmatic than Harriot's, and he went on to explain what to do when, as must often happen in practice, the second differences are not exactly equal (he suggested taking an average). The main similarity with Harriot's calculations is in the layout, as can be seen by comparing Tables 23 and 24 with Tables 11 and 12; the right alignment in the middle column, used by both Briggs and Harriot, aids the arithmetic to some extent, but is not seen in later writers, who generally give first differences in a new column further to the right.

The question of whether Briggs and Harriot knew each other, or in any way shared their mathematical knowledge, remains tantalisingly unanswered. We know that both were deeply interested in the mathematics of navigation, and also with the construction and interpolation of tables, and there can be no doubt that if they had ever met they would have had much to talk about, but there is no evidence that they did so. From 1596 to 1620 Briggs was based at Gresham College as Professor of Geometry, while Harriot was some way out of London at Syon House at Isleworth, living and working under private patronage and surrounded by his own circle of acquaintances. It is therefore quite possible that even if they had heard of one another they never met in person.

It is clear that Torporley and Briggs, on the other hand, did know one another, but Torporley had lived abroad until 1603 and then in relative isolation in Worcestershire from 1608 until at least 1622. Did he meet Briggs between 1603 and 1608 while Harriot was still alive, or only some years later, after his death? By 1625 Briggs certainly knew of the efforts to edit Harriot's papers because he informed Kepler that publication was expected very soon.[60] Then in 1630, George Hakewill, in the second edition of *An apologie*, published a letter from Briggs listing 'the most observable inventions of moderne Mathematicians unknowne to the Ancients', and one of them, according to Briggs, was the understanding of spherical triangles, which he claimed was first taught by Harriot.[61] Harriot's work on spherical triangles was never published, and the fact that Briggs wrote about it and considered it 'modern' suggests that he had heard about it relatively recently (in the last ten years, say, rather than the last thirty; he mentioned it *after* Napier's logarithms, for instance). In that case he may well have learned of it from Torporley between 1622 and 1625. Could the same be true of Harriot's method of differences?

It is at least conceivable that Torporley in 1622 or 1623 demonstrated Harriot's difference method to Briggs. It is also likely, however, that by then Briggs had already worked out a version of the method for himself. Gellibrand in his brief preface to Briggs' *Trigonometria britannica* in 1633, asserted that Briggs had constructed his tables using algebraic equations and differences around thirty years before, though there

[60] 'Cum propediem expectemus et exoptemus ipsius auctoris librum posthumum,' ('Since we may expect and hope for a posthumous book from that author any day'), Briggs to Kepler, 10 March 1625, in Kepler 1858–72, IV, 661–662.

[61] 'primus docuit peritissimus Geometra Thomas Hariottus, cum ante eum nemo hoc sit assequutus.' ('[which] the most skilled geometer Thomas Harriot first taught, since before him no-one had pursued this.') Hakewill 1630, 263–264. Briggs was wrong on this point: Viète had taught many rules for both plane and spherical triangles in his *Variorum responsorum Liber VIII* of 1593, a book that Harriot had studied in detail, but Viète's treatise was rare and it is likely that Briggs had never seen it.

is no independent evidence for such a claim, which may be somewhat exaggerated. We simply do not know how or when Briggs devised his various interpolation methods, or which of them he used in 1617 to construct his first chiliad, later presented as a gift to Torporley. All we can say with certainty is that Harriot was using constant difference interpolation methods at least ten years before an example appeared in print in Briggs' *Arithmetica logarithmica* in 1624.

Walter Warner (*c.* 1557–1643)

Walter Warner was employed from about 1590 as keeper of the books and scientific instruments of Henry Percy, the ninth Earl of Northumberland; after 1617 he continued to hold a small pension from the Earl. Harriot also lived under Percy's patronage from the mid 1590s, and so knew Warner well, and named him in his will as one of the men Torporley should consult if anything in his later work was unclear. When the task of editing Harriot's papers was divided up between Torporley and Warner some time in the early 1620s, Torporley handled Harriot's method of differences and also his work on Pythagorean triples.[62] It fell to Warner to edit Harriot's work on equations, and in 1631 he produced the *Artis analyticae praxis*, the only publication of any of Harriot's mathematics until modern times. We might now have had clearer expositions both of the method of differences and of Harriot's work on equations if the division had been made the other way round: Torporley understood Harriot's theory of equations much better than Warner did,[63] but Warner eventually provided clearer expositions of the method of differences.

Warner almost certainly learned something of the interpolation method during the years when Harriot was using it most actively, between about 1610 and 1614. He was less frequently in contact with Harriot when the 'Magisteria' was written in its present form in 1618 or later, but he knew many of the formulae that appeared in it, for amongst his manuscripts are several pages of notes and calculations relating to them.[64] The notes are undated but they are in a firm hand, which suggests that Warner made them well before 1630, and possibly before 1620 while Harriot was still alive.[65]

In view of Warner's later influence, we need to look more closely at some particular pages from his manuscripts in which he explored the method of differences. For convenience we have labelled the items as (W1) to (W5).

(W1) Two sheets headed 'Problema Arithmeticum ad doctrinam de differentium Progressionibus pertinens'.

Now CUL Add MS 9597/9/15, ff. 72–73.

(W2) Ten sheets beginning 'Prob. 1. Pro tabularia inveniendo in serie crescente'.

Now BL Add MS 4396, ff. 20–29.

[62] See Tanner 1977.

[63] See Stedall 2003, 22–24.

[64] BL Add MSS 4394, ff. 350–372 and 4396, ff. 20–29.

[65] By the late 1620s Warner's handwriting had become very shaky, see, for example, some draft material for the *Praxis* in BL Add MS 4395, f. 92.

(W3) A sheet beginning 'Methodus calculatoria, qua canon analogicus …'

Now BL Add MS 4396, f. 19 (with associated rough working on f. 18).

(W4) A method for solving polynomial equations using difference tables.

Now BL Add MS 4395, ff. 166–166v and 181.

(W5) A list of polynomials, evaluated for the first few positive integers, with successive differences.

Now BL Add MS 4394, f. 359.

Here the contents of each will be described briefly. Warner's mathematical work beyond the *Praxis* has never previously been given any attention, so this short account offers a new insight into his mathematical interests in the later years of his life.

(W1) is headed 'Problema Arithmeticum ad doctrinam de differentium Progressionibus pertinens' (An arithmetic problem pertaining to the doctrine of progressions of differences), and gives the algebraic difference table shown in Table 25.

$$
\begin{array}{rl}
b = & b \\
& b + 1a \\
& b + 2a + e \\
& b + 3a + 3e \\
& b + 4a + 6e \\
& b + 5a + 10e \\
& b + 6a + 15e \\
& b + 7a + 21e \\
& b + 8a + 28e \\
& b + 9a + 36e \\
c = & b + 10a + 45e \\
& b + 11a + 55e \\
& \dots \\
& \dots \\
& \dots \\
d = & b + 20a + 190e
\end{array}
$$

Table 25. Difference table from Warner's 'Problema arithmeticum'.

In some notes alongside the table Warner explained that he would denote the difference between b and c by f, the difference between c and d by h, and the difference between f and h by g. We may rewrite the first two relationships as

$$
\begin{aligned}
f &= c - b = 10a + 45e \\
h &= d - c = 10a + 145e
\end{aligned}
$$

from which it is easy to see that

$$g = h - f = 100e.$$

It therefore follows that $e = \frac{g}{100}$, and also, from the first equation, that $a = \frac{f-45e}{10}$. Warner gave a rather longer derivation but arrived at the same formulae for e and a.

(W2) is a run of ten sheets (ff. 20–29) concerned with the following two problems for a constant difference table: (1) given any index $\frac{N}{n}$ in the margin, find the corresponding entry in the table; (2) conversely, given an entry in the table find its index $\frac{N}{n}$. Both problems were addressed by Harriot at the end of the 'Magisteria', on page 35 (for constant first differences) and on page 36 (for constant second differences).

The first four pages in the present ordering of the manuscripts (ff. 20–23) investigate the problem for constant second differences, and offer four worked examples in the form of problems; for convenience we may denote these pages by (W2a).

Problem 1 asks for the entry corresponding to $\frac{N}{n} = \frac{3}{5}$, that is, the third of four new terms (or five spaces), interpolated between 3 and 48 in Table 26.

N	C		
		B	
0	3		A
		45	
1	48		50
		95	
2	143		50
		145	
3	288		50
		195	
4	483		50
		245	
5	728		

Table 26. A difference table to be interpolated, from Warner's 'Problem 1'.

Warner wrote $C = 3$, $B = 45$, $A = 50$, and then used Harriot's formula

$$C + \frac{2.NnB - NnA + NNA}{2.nn}$$

to calculate that the required term is 24. This problem is identical to that on page 36 of the 'Magisteria', where Harriot showed that the first few terms of the interpolated table are as shown in Table 27 (Warner also gave this table).

$\frac{N}{n}$	c	b	a
0	3		
		5	
$\frac{1}{5}$	8		2
		7	
$\frac{2}{5}$	15		2
		9	
$\frac{3}{5}$	24		2
		11	
$\frac{4}{5}$	35		2
		13	
$\frac{5}{5}$	48		

Table 27. The interpolated difference table, from the 'Magisteria' (page 36).

Problem 2 is similar, but for decreasing instead of increasing columns. Problems 3 and 4 address the converse question: given a number in the table find the corresponding index $\frac{N}{n}$ in the margin, for increasing and decreasing columns respectively.

These four problems are followed by six pages (ff. 24–29) in which Warner explored similar questions for general tables with constant second difference; we denote these six pages by (W2b). They contain two problems (ff. 24–25), two theorems (ff. 26–27), and two worked examples relating to the theorems (ff. 28–29).

(W3) comes immediately before (W2) in the present ordering of the manuscripts. It is placed at the end of a short treatise on the construction of tables of antilogarithms, or as Warner and others in the seventeenth century called them, 'analogics'.[66] Warner devoted the final years of his life to the computation of such tables, an ambitious undertaking, and in some ways a futile one, because for all practical purposes ordinary tables of logarithms could just as well be read in reverse. Antilogarithms need be calculated only for numbers between 0 and 1 since all others are easily derived from these by multiplying by an appropriate power of 10. According to Pell fifty years later, Warner began with a table of just 10 entries, for 0.1, 0.2, …, 0.9, 1.0, which he called his first canon, and then interpolated nine new entries between each existing pair to give the second canon, a table of 100 entries, for 0.01, 0.02, …, 0.99, 1.00. Repeating the process gave him his third canon, of 1000 entries, and a fourth, of 10,000 entries.[67] Pell's assertions may have been no more than conjectures made long after the event. Accurate interpolation into the first, second, or third canons would have required calculation to a very large number of decimal places, and there is no evidence

[66] BL Add MS 4396, ff. 1–17.

[67] Information from notes made by Pell in 1683, BL Add MS 4424, ff. 1–3.

of such calculations in Warner's surviving manuscripts. We do know, however, that by 1639 Warner was in possession of a completed table of 10,000 entries, each calculated to twelve decimal places, and that he was ready to carry out a new round of interpolation to create 100,000 entries, each to ten decimal places. The examples in (W3) show how this process is carried out, and the page begins as follows:

> *Methodus calculatoria, qua canon analogicus ex decem terminorum proportionalium millibus iam confectus ad centum millia promovendus fit per insertionem novem terminorum inter singulos binos totius canonis. Exemplificata in duobus primis et in duobus ultimis.*
>
> A method of calculation, by which the analogic canon now composed of ten thousand proportional terms may be made up to one hundred thousand by the insertion of nine terms between each pair of the whole canon. Exemplified by the two first and the two last.

Warner's first example shows how to interpolate nine new entries at the beginning of the table, between 0 and 0.00010, namely, those for 0.00001, 0.00002, ..., 0.00009. Over this range the second difference (taken as far as the twelfth decimal place) can be considered constant, and Warner's method is exactly Harriot's interpolation method for constant second differences. His second example, on the reverse of the same sheet, gives a similar calculation for the end of the table, for the entries to be interpolated between 0.99990 and 1.

(W4) is part of a long and somewhat rambling exposition by Warner on a number of mathematical subjects, including quadratic and cubic equations, the summation of geometric series, and properties of the hyperbola.[68] After an obscure and wordy discussion of cubic equations, Warner began a new page (numbered 21 in his sequence) with the following words:[69]

> There is another way for the solution of equations by differentiall progressions to be considered of. As for squares
> $aa + 3a = 130 \quad // \quad a = 10$

He then gave Table 28, which shows values of $aa + 3a$ for $a = 1, 2, 3, 4$.

$$
\begin{array}{lllllll}
a = 1 & // & \equiv 4 & --- & 4 & --- & 2 \\
a = 2 & // & \equiv 10 & --- & 6 & --- & 2 \\
a = 3 & // & \equiv 18 & --- & 8 & --- & 2 \\
a = 4 & // & \equiv 28 & --- & 10 & --- & 2
\end{array}
$$

&c. usque ad 130. [etc. as far as 130.]

Table 28. Warner's table of values for $aa + 3a$.

[68] BL Add MS 4395, ff. 155–182.
[69] BL Add MS 4395, f. 166.

Clearly this is very similar in construction to Harriot's table for $1C + 3R$ (see Table 13). It is followed in Warner's manuscript by a similar table for a cubic equation, $aaa + 2aa + 3a = 1230$; see Table 29. This example was also given by Harriot (see the second row of Table 16). Continued far enough, the table would show that $a = 10$ is a root of the given equation.

$a = 1$ //	$=\!=$ 6	$---$	6	$---$	4	$---$	6
$a = 2$ //	$=\!=$ 22	$---$	16	$---$	10	$---$	6
$a = 3$ //	$=\!=$ 54	$---$	32	$---$	16	$---$	6
$a = 4$ //	$=\!=$ 108	$---$	54	$---$	22	$---$	6
$a = 5$ //	$=\!=$ 190	$---$	82	$---$	28	$---$	6

&c. usque ad solidum datum. [etc. as far as the given 'solid', namely, 1230.]

Table 29. Warner's table of values for $aaa + 2aa + 3a$.

A third table gives values of a fourth degree polynomial, $aaaa + 1aaa + 3aa + 2a$, and now the constant difference is 24, in the fifth column.

All Warner's tables are set out in neat rectangular blocks as though derived from the first row, $a = 1$. As we saw in discussing Harriot's tables, this first row together with the constant difference in the final column makes it easy to continue such tables as far as required. It is possible to find the first row by extrapolating upwards from the first few calculated values, as Harriot seems to have done in some of his early explorations (see Tables 13 and 15), but Warner also gave rules for writing it down immediately:[70]

> The primes [first terms] of these progressions are given thus first of the squares, the first is the summe of the coefficients, the second is the same, the third the characteristic 2. Of the cubes the first is the summe of the coefficients, the second the same, the third the double of the coefficients of the squares, the fourth the characteristic 6. Of the biquadrates the first the summe of the coefficients, the second the same, the third the double of the coefficient of the squares more by 2, the fourth the sextuple of the coefficient of the cubes less by 12, the fifth the characteristic 24, the like rules are to be found for the rest [...]

Warner gave no hint as to how he knew these rules but, as suggested earlier, they are not hard to find by inspection (see Tables 16 and 32). It could well be that Warner learned them directly from Harriot who, as we have seen, wrote out several tables of the same kind.

Fifteen folios later, Warner added a further note on solving equations, referring back to the material given earlier.[71] Now he showed how to deal with a case where the root is not an integer. His example is based on the equation $aa + 6a = \dots$, for which he wrote out Table 30. The sign $=\!=$ in this table is not an equals sign but may be read as 'gives' or 'yields'.

[70] BL Add MS 4395, ff. 166–166v.

[71] 'This note is to be referred to the 21. page.' BL Add MS 4395, f. 181.

1	=	7	--	7	--	2
2	=	16	--	9	--	2
3	=	27	--	11	--	2
4	=	40	--	13	--	2
5	=	55	--	15	--	2

Table 30. Warner's table of values for $aa + 6a$.

Now suppose the required value falls between 7 and 16, that is, the root lies between 1 and 2. Such intermediate values can be found by interpolation, as Warner explained:

> Any of the spaces of this progression [that is, 7, 16, 27, ...] may be subdivided into ten parts through only adding $\frac{9}{10}$ to the first of the second series, which holds generally where the constant difference is 2 as in all resolutions of the square equation it is.

A new table (Table 31) shows what he meant: the first entry in the penultimate column is increased from 7 to $\frac{79}{10}$ and the rest of the table is filled in simply by adding differences. Observe Warner's use of Harriot's notation $\frac{11}{10}, \frac{12}{10}, \ldots$ to indicate interpolations between the first and second entries from the original table. These are not fractions, but indices, and now the $=$ sign is to be interpreted loosely as 'is'. The interpolated values themselves (after the $=$ signs) *are* fractions. In the second and third columns, where he wrote denominators at all, Warner wrote them as 10 and 1 where we might expect them all to be 100, probably because they were intended to be values of $f = 10a$ and $g = 100e$, respectively. Most of the time, however, he simply omitted them.

$\frac{10}{10}$	=	$\frac{700}{100}$	--	$\frac{79}{10}$	--	$\frac{2}{1}$
$\frac{11}{10}$	=	$\frac{781}{100}$	--	$\frac{81}{}$	--	$\frac{2}{}$
$\frac{12}{10}$	=	$\frac{864}{100}$	--	$\frac{83}{}$	--	$\frac{2}{}$
$\frac{13}{10}$	=	$\frac{949}{100}$	--	$\frac{85}{}$	--	$\frac{2}{}$
...						
...						
...						
$\frac{20}{10}$	=	$\frac{1600}{100}$	--	$\frac{99}{10}$	--	$\frac{2}{1}$

Table 31. Warner's interpolation of Table 30.

Once again Warner has given an easy rule (add $\frac{9}{10}$ to the leading first difference) but without explanation, suggesting that he, or perhaps Harriot, had already worked a number of examples of this kind.

This afterthought on equations is inserted into what is otherwise a discussion of properties of the hyperbola. The discussion comes to an end on the folio following Table 31, with nothing else on interpolation.

(W5) gives a list of polynomials, each evaluated for the first few positive integers, and with successive differences shown alongside. Five of the examples are shown in Table 32. In each case the entries in the first column are values of the given polynomial for $a = 1, 2, 3, \ldots$, and the subsequent columns contain successive differences.[72] These are clearly similar to the tables in (W4).

$aa + 3.a$	$=\!=$	4.	4.	2		
		10.	6.	2		
		18.	8.	2		
$aa + 5.a$	$=\!=$	6.	6.	2		
		14.	8.	2		
		24.	10.	2		
$aaa + 4.a$	$=\!=$	5.	5.	0.	6	
		16.	11.	6.	6	
		39.	23.	12.	6	
		80.	41.	18.	6	
$aaa + 2.aa$	$=\!=$	3.	3.	4.	6	
		16.	13.	10.	6	
		45.	29.	16.	6	
		96.	51.	22.	6	
$aaaa + 3.a$	$=\!=$	4.	4.	2.	−12.	24
		22.	18.	14.	12.	24
		90.	68.	50.	36.	24
		268.	178.	110.	60.	24
		640.	372.	194.	84.	24

Table 32. Polynomials evaluated for $a = 1, 2, 3, \ldots$, with successive differences.

Unlike (W4), Table 32 is not part of a more general discussion of equations, but falls amongst working that is clearly related to the 'Magisteria', including algebraic 'binomial coefficients', difference tables, and so on. If Warner and Harriot worked on this material together while Harriot was still alive one would expect to find more

[72]BL Add MS 4394, f. 359.

of it amongst Harriot's manuscripts, but there seems to be very little. It is just possible, however, that Harriot's notes on it were among the sheets headed 'Poristicum arithmeticarum progressionum', which Torporley in 1627 described as lost.

A further small piece of evidence concerning Warner's interest in solving equations by differences is to be found in Hartlib's diaries, where an entry for 1635 notes:[73]

> Mr Warner hase all Hariots MS and is setting some of them forth. There is also a new and better kind of Logarithms coming forth then ever yet hase beene knowen, which principally will serve for the Algebraical Operations.

The 'new and better kind of Logarithms' would have been Warner's tables of antilogarithms, while 'Algebraical operations' could certainly be taken to mean 'solving equations'; indeed it is hard to see what else it could mean. Here already perhaps are the seeds of the rumour that Collins repeated later: that the primary purpose of Warner's antilogarithms was in fact for solving equations.[74]

Having seen some of the mathematical work that occupied Warner during the 1630s, we turn now to the two men with whom he shared it most closely, Sir Charles Cavendish and John Pell.

Charles Cavendish (*c.* 1595–1654)

Though he never displayed any great skill in mathematics, Sir Charles Cavendish took a keen interest in it from the 1620s onwards. He knew about the publication of Harriot's manuscripts because he gave a copy of the *Praxis* to Robert Payne, chaplain to his brother William, in December 1631.[75] Cavendish would almost certainly have sought out the editors, and by 1634, if not earlier, was acquainted with Aylesbury and Warner.

As noted right at the beginning of this essay, Cavendish wrote of the 'Magisteria' that he was 'so farre in loue with it that I coppied it out', and there is indeed a complete copy in his surviving manuscripts.[76] There are also about twenty other sheets containing difference tables and interpolation formulae.[77] Occasionally these contain direct references to pages of the 'Magisteria' while others mention 'Mr Warner'. Of particular interest are the following:

(C1) A copy of (W1), Warner's 'Problema arithmeticum'. Cavendish has added 'Mr: Warner' at the end.

Now BL Harleian MS 6083, ff. 449–450.

On the reverse of f. 449 (and f. 448) is a worked example copied from Harriot's 'Of unaequall progression of sines'. Cavendish copied the table shown in Table 9 (including the triangles and the asterisk), together with the associated working for finding the 'sine' of 37″.

[73] Hartlib, *Ephemerides*, 29/3/41A.

[74] Rigaud 1841, i, 215–216; II, 219.

[75] This copy was later deposited in the Bodleian Library, Oxford, shelfmark Savile O.9.

[76] Cavendish's copy of the 'Magisteria' is in BL Harleian MS 6083, ff. 403–444; many of these sheets, especially near the beginning, have related rough working on the reverse side.

[77] BL Harleian MS 6083, ff. 27–28, 61–66, 190–193, 445–455.

(C2) A copy of the four problems given by Warner in (W2a).

Now BL Harleian MS 6083, ff. 445–448.

Brief versions of these problems are also copied elsewhere in Cavendish's manuscripts, always attributed to Warner.[78]

These copies of Warner's material were probably made during the 1630s, when Cavendish and Warner were most closely in contact. In 1644, the year after Warner died, Cavendish went into exile on the continent, and would have had no further opportunity to peruse the papers, which remained in England with Warner's nephew. In Cavendish's manuscripts, the material from Warner comes immediately after the 'Magisteria', and contains a page reference back to it, suggesting that all the copying was done around the same time. Thus when Cavendish in 1651 wrote of the 'Magisteria' that he had 'coppied it out', it is likely that he was referring to a copy he had made some time in the 1630s.

Cavendish's letter was addressed to John Pell, whom he had met in or shortly before 1640.[79] For over ten years, Pell had acted as Cavendish's mathematical adviser, and Cavendish frequently asked him to explain passages that were unclear to him in the various mathematical texts he tried to read. In 1643 Pell moved to the Netherlands, where he taught first in Amsterdam and then in Breda, while Cavendish led a peripatetic existence between Hamburg, Paris, and Antwerp, but they continued to correspond on a variety of mathematical and scientific subjects.[80] The letter in which Cavendish asked for explanation of Harriot's 'doctrine' was almost the last in their exchange. The request was similar to many that Cavendish had made over the years, but this time it was made on behalf of a third person, Thomas Aylesbury: 'Sr. Th. Alesburie remembers him to you & desires to knowe if you would be pleased to shew the vse of Mr: Harriots doctrine of triangulare numbers; which if you will do he will send you the original.'

Aylesbury had left England in 1649 and met Cavendish in Antwerp in 1651. Cavendish's letter is remarkable evidence that Aylesbury had brought some of Harriot's manuscripts with him into exile. Perhaps all of them: there is nothing to suggest that the 'Magisteria' was ever separated or treated differently from the others amongst which it now rests. The letter from Cavendish is the last surviving reference to the original manuscript of the 'Magisteria', whose fate from then until the late eighteenth century is unknown. Aylesbury died in 1657 and it seems that he or his executors returned Harriot's manuscripts, as Harriot had long since requested, to the library of the Earl of Northumberland. The Royal Society instigated searches for them amongst Aylesbury's papers in the 1660s but failed to find them, and the generally held view in the later seventeenth century was that they were lost. A century later they were finally rediscovered at Petworth, the 'Magisteria' among them.

[78]BL Harleian MS 6083, f. 27 and f. 65.

[79]Pell wrote out some mathematical solutions for Cavendish in June 1640; see BL Add MS 4431, f. 270v.

[80]The full correspondence is to be found in Malcolm and Stedall 2005, 335–586.

John Pell (1611–1685)

Antilogarithms. As we have seen, some time in the 1630s Walter Warner embarked on the formidable task of constructing a very large table of antilogarithms, and by 1639 had begun on the final round of interpolation that would take him from 10,000 to 100,000 entries. In November 1639 he was fortunate enough to be introduced to John Pell, then aged twenty-eight and living in London but out of work.[81] Pell himself had always been intrigued by the construction of tables: as a student at Cambridge in 1628 he had already written to Briggs with queries about interpolating tables of sines and logarithms.[82] He seems to have taken an immediate interest in Warner's project, because five days after their first discussion Warner showed him the first 10,000 entries of the extended table. The work was laborious, however, and Warner was getting old. Eventually, in 1641, Warner offered Pell £40 to complete the tables and make them ready for printing.[83] Pell began the task in June 1641 and in order to account to Warner for his time kept an extraordinarily detailed account of his activities.[84] For 25 June, for example, his first day of calculation, his activities hour by hour were as follows:

25 Friday
1h I drew ye black lead perpendiculars in 5 calculatory pages
I drew ye Inky transposall lines in 5 calculatory pages

I transcribed into those 5 pages all ye 50 analogicalls which lead ye work (yeir first differences wanting but 10).

2h I calculated all ye 2nd differences ($g = 100e$)
subtracted their tenths
bisected yeir remainders 45*e*
subtracted those halfes from both *f* and *h* to find ye upper and lower 10*a*
Tried ym by adding *g* to ye lower 10*a*
Found and wrote down *a* and *e* in every semi-column of ye
500 in a little lesse yn an houre
3h I calculated 90 and ye differences of ye last 10 in one houre
4h I calculated 80 and ye differences of ye other 20 in one houre
5h I calculated a round century
6h I calculated 90 and ye differences of ye last 10 in one houre
7h calcul [*sic*] all ye differences and all ye numbers save 12
8h transcribed 250 } 500 in 2 houres just
9h transcribed 250 }

[81]See Malcolm and Stedall 2005, 84–85.

[82]Pell's letter to Briggs is lost, but we have Briggs' reply in BL Add MS 4398, f. 137, and printed in Halliwell-Phillips 1841, 55–57.

[83]The sum of £40 would be worth about £5000 or 10000 US dollars in modern currency.

[84]BL Add MS 4365, ff. 36–39. A table similar to Table 33 appears at f. 40.

From these and other entries it is possible to reconstruct his method. It is most easily described with the help of Table 33, from which it can be seen that Pell's notation is precisely that suggested by Warner in (W1).

b		
	$f = 10a + 45e$ ('lower $10a$' $+ 45e$)	
$b + 10a + 45e$		$g = 100e$
	$h = 10a + 145e$ ('upper $10a$' $+ 45e$)	
$b + 20a + 190e$		

Table 33. A reconstruction of Pell's method.

Once he had calculated first and second differences, Pell proceeded as follows. First he subtracted one tenth of g from g itself (to give $90e$) and half of this gave him $45e$. Subtracting $45e$ from h and f gave him what he called 'upper $10a$' (actually $10a + 100e$) or 'lower $10a$' ($10a$ itself). The difference between 'upper $10a$' and 'lower $10a$' is $100e$, or g, which provided a useful check on his working. Dividing g by 100 gives e, and dividing 'lower $10a$' by 10 gives a. This was all that was needed to calculate the nine intermediate values, as listed in Table 25.

By the end of August 1641, Pell was beginning to complain about the cost of the work and Warner asked for all the papers back. Pell's careful inventory of returned items includes 'his Problema Arithmeticum in 2 folios',[85] almost certainly (W1). The difficulties were resolved shortly afterwards, and Pell continued to work on the tables, which were completed and sent to Warner in July 1642.[86] Unfortunately, the outbreak of civil war prevented their publication and Warner died the following year. In 1650, the tables, along with Warner's other mathematical papers, were given by his nephew to Herbert Thorndike, who tried to persuade Pell to publish them, but without success. After Thorndike's death in 1672 the papers went to Richard Busby of Westminster School, and Pell looked over them again at Westminster in 1683, two years before he died.[87] Reams of papers owned by Busby were later deposited in the British Museum,[88] but Pell's antilogarithm tables were not amongst them, and have never been found.

One might wonder yet again why Warner and Pell thought it worthwhile to engage in such enormous labour, but there is just one hint that Harriot himself had proposed such a table. This was reported by John Wallis, who in the 1693 Latin edition of his *Treatise of algebra* wrote:[89]

---

[85] BL Add MS 4365, f. 37v.

[86] For further details see Malcolm and Stedall 2005, 92–93 and 280–286.

[87] BL Add MS 4424, ff. 1–3 and 10.

[88] Busby inherited the mathematical papers of both Warner and Pell, now BL Add MSS 4394–4404 and 4407–4431.

[89] 'Canonem illum nescio an inchoaverit D. *Thomas Harriot*; cujus Schediasmata nactus D. *Walterus Warner*, inde edidit ejus Algebram, Anno 1631; & spem fecit ejus se plura editurum. Idemque *Warnerus* non

> I do not know but that such a canon was begun by Mr Thomas Harriot, whose papers Mr Walter Warner obtained, from which he edited his Algebra in the year 1631, and promised that he would edit more of his. Likewise Warner not much later (if he did not also himself first begin it) completed that canon and prepared it for the press; now, I believe, around fifty years ago (if not more). Which was recently disclosed to me by Mr John Pell, who was intimately acquainted with Warner; and helped him, having assisted him in part of the work of calculation.

Pell gave Wallis this information too late for inclusion in the first edition of his *Treatise of algebra* which went to press in 1684; probably he told Wallis about the tables some time after he looked over them again at Westminster in 1683. As far as the bare facts are concerned, Wallis's report is substantially correct, but it is hard to tell whether the suggestion about Harriot's involvement reflects something Wallis had learned from Pell, or was merely a speculation of his own, another useful weapon in his struggle to promote Harriot's reputation. All we can say is that the methods used by Pell and Warner, if not the tables themselves, most certainly came from Harriot.

Pell continued throughout his life to be interested in constant difference methods for the interpolation of tables. After he became acquainted with John Collins, for example, probably in the early 1660s, he discovered that Collins owned one of the few copies of Briggs' *Arithmetica logarithmica* containing three extra sheets, with logarithms from 100,000 to 101,000, and he made use of these in conjunction with difference methods to calculate additional logarithms.[90]

As Harriot had done, and presumably Warner also, Pell always took differences in the positive direction; in other words, he worked with what we would now call absolute differences. For an increasing table of antilogarithms, the absolute differences in each column increase, but in an increasing table of logarithms the absolute differences decrease in every column after the first.[91] To apply a constant difference method, however, it has to be assumed that the differences are eventually constant. Pell wrote out a lengthy difference table of this second kind,[92] with decreasing first and second differences, running to 60 rows. A small section of it is shown in Table 34.

---

ita multo post (si non & ipse primus inchoaverit, saltem) Canonem illum absolvit, & prelo paravit; jam ante annos, credo, Quinquaginta circiter (si non & plures.) Quod mihi nuper indicavit D. *Johannes Pell*; qui & *Warnero* fuerat familiariter notus; eique fuerat auxilio, parteis laboris Calculi subeundo.' Wallis 1693–99, II, 63.

[90]BL Add MS 4413, ff. 229–231; for a worked example see Malcolm and Stedall 2005, 303–305. Collins' copy of the *Arithmetica logarithmica* was held in the library of the Earls of Macclesfield until 2004 when it was sold by auction at Sotheby's of London to a private buyer.

[91]The derivatives of antilog x are multiples (depending only on the logarithmic base) of antilog x, and so *increase* in value as x increases; the derivatives of $\log x$, on the other hand, are multiples of $\frac{1}{x}$, $-\frac{1}{x^2}$, $\frac{1}{x^3}$, ..., all of which *decrease* in absolute value as x increases.

[92]BL Add MS 4415, f. 113.

$$
\begin{array}{llll}
\dots & & & \\
\dots & & & \\
\dots & & & \\
A-3b-6c-10d & & & \\
 & b+3c+6d & & \\
A-2b-3c-4d & & c+3d & \\
 & b+2c+3d & & d \\
A-b-c-d & & c+2d & \\
 & b+c+d & & d \\
A\mp 0b-0c-0d & & c+d & \\
 & b\pm 0c+0d & & d \\
A+b-0c\mp 0d & & c & \\
 & b-c+0d & & d \\
A+2b-c+0d & & c-d & \\
 & b-2c+d & & d \\
A+3b-3c+d & & c-2d & \\
 & b-3c+3d & & \\
A+4b-6c+4d & & & \\
\dots & & & \\
\dots & & & \\
\dots & & &
\end{array}
$$

Table 34. One of Pell's difference tables, with decreasing first and second differences.

Equations. Pell also experimented with difference methods in another branch of mathematics that greatly interested him: solving polynomial equations. As we have already seen, polynomial expressions evaluated at regular intervals produce constant difference tables, so constant difference interpolation is a natural way of solving polynomial equations. Pell knew this, but could never be prevailed upon to explain the method, and for the most part we have only second hand information from Collins, who was never quite sure what was involved. Collins was in particularly close contact with Pell after July 1669 when Pell boarded with him for several months, but they must have discussed equation solving before then, because from 1667 onwards Collins repeatedly asserted that Pell could solve equations using tables.[93] Collins was confused as to whether the necessary tables were of sines, logarithms, or antilogarithms, which suggests that he may have simply failed to distinguish different techniques. He himself published a paper on solving equations, in the *Philosophical Transactions* of April 1669, and commented that this work was part of a longer narrative 'formerly made' by himself 'touching some late Improvements of Algebra in England'. Some manuscript notes in Collins' hand on 'Improvements of Algebra in England' were therefore probably written during 1668 or early in 1669.[94] In them he wrote that 'a learned person' had made tables of progressional differences and that the use of such tables was for 'the easy finding of the rootes'. There is little doubt that the 'learned person' was Pell, in

[93]In chronological order, Collins' references are: Rigaud 1841, II, 472–473; II, 197; II, 219–220; I, 215–216; I, 243, 247.

[94]BL Add MS 4474, ff. 1v–4.

whose writings many such tables survive (see, for example, Table 34). Collins went on to say that the method seemed to consist of 'interpoling [sic] such rankes whose 3^{d} 4th 5th, 6th Differences are aequall'.

Collins claimed that Pell's use of tables for solving equations went back to the early 1640s, in other words, to the time when he was working with Warner. In December 1670 he wrote to James Gregory that: 'Dr Pell affirmeth he hath for above 30 yeares used to solve Aequations by tables, …'. As usual Collins' description of the method was unenlightening, and Gregory regretted that he had not 'explicat more fully',[95] but it is likely that Collins' historical understanding was correct and that Pell's method had indeed arisen from his work with Warner thirty years earlier.

In the 1670s Collins penned an account of Pell's achievements for Leibniz, and after describing one of Pell's tables (a yard long, according to Collins) and its uses, he made the following remark:[96]

> … to attempt the same [solving equations] in Vieta's method, Mr Warner used to call work unfit for a Christian, and more proper to one that can undertake to remove the Italian Alps into England.

Viète in his *De potestatum … resolutione* of 1600 had taught a method of numerical solution by a process of successive approximations. Harriot had explored the method at some length,[97] and Warner had reproduced a few examples of it in the *Praxis*. In calling it work 'unfit for a Christian' did he simply mean that it was long and difficult? Or was he comparing it with another method, in his view much better: solution by tables?

In 1673 Leibniz visited London from Paris, and in conversation with Pell remarked that he had found a method of interpolating series using differences. Pell replied, to Leibniz's embarrassment, that such methods were already well known even in France, and had appeared in Gabriel Mouton's *Observationes diametrorum solis et lunae apparentium* in 1670. Pell might have embarrassed Leibniz even further if he had told the whole truth: that such methods were known to Harriot well before 1620, and that he himself had used them extensively for over thirty years.[98]

Nicolaus Mercator (1619–1687)

We now return to the 'small quarto' by Nicolaus Mercator, mentioned by William Jones to Roger Cotes in 1713, and loaned to him alongside Torporley's 'Congestor'.[99] After Jones' death both documents were kept in the library of the Earls of Macclesfield, until they were transferred to Cambridge University Library in 2000. Mercator's treatise is bound, like many of the mathematical items from the Macclesfield library, in a blue paper cover, and on the flyleaf is written in a later hand 'A MS of Mr Nicolaus Mercator,

[95] Gregory 1939, 142, 169.

[96] Rigaud 1841, I, 248.

[97] See Stedall 2003, 45–123.

[98] See Hofmann 1974, 26–29; the account given there of Leibniz's method suggests that it was very similar indeed to Harriot's.

[99] Now CUL Add MS 9597/9/15.

containing Theorems relating to the Resolution of Equations, The method of Differences, and the Construction of Tables'. In fact the manuscript is not Mercator's original but a copy in the hand of John Collins. On the first page it is headed 'The Doctrine of Differences – by N: M:', and for convenience we will use this title in what follows.

Mercator's 'Doctrine of differences' consists of 38 pages, the first 14 of which are about general methods of solving quadratic and cubic equations. There is then a section of four pages on solving equations by difference methods, and Mercator's examples turn out to be precisely those given by Warner in (W4). His first was the quadratic equation $aa + 3a = 130$, with $aa + 3a$ evaluated for $a = 1, 2, 3, 4$ (compare Tables 35 and 28).[100] '*Homog.*' is 'the homogene of comparison', the term free of a, while 1^{ae} and 2^{dae} are abbreviations for *primae* and *secundae*.

		Homog.		*diff* 1^{ae}		*diff* 2^{dae}		
$a = 1$	//	$= 4$	––	4	––	2	}	
$a = 2$	//	$= 10$	––	6	––	2	}	*&c. usque ad* 130.
$a = 3$	//	$= 18$	––	8	––	2	}	[etc. as far as 130.]
$a = 4$	//	$= 28$	––	10	––	2	}	

Table 35. Mercator's table of values for $aa + 3a$.

Mercator then observed that other values of $aa + 3a$ can be found by interpolating between the values in the first progression (after the marks //), and he quoted Warner (W4) exactly: 'Any of the Spaces of this Progression', he wrote, 'may be Subdivided into ten parts through only adding $\frac{9}{10}$ to the first of the second Series which holds generally where the constant difference is 2; as in all resolutions of the Square Equation it is.' As an example he offered the equation $aa + 6a =$ 'a number', and gave the tables from (W4) (Tables 30 and 31), demonstrating 9 interpolations between the values 7 and 16.

Mercator went on to give Warner's other examples of polynomial values, with the tables for $aaa + 2aa + 3a = 1230$ (see Table 29), and for $aaaa + 1aaa + 3aa + 2a =$ 'a number', both from (W4). He also gave the rules for constructing the crucial first row:

1. First of the squares
The first is the sum of the coefficients
The second is the same
The third is the Characteristick 2

2. Of the cubes
The first is the sum of the coefficients
The second is the same
The third is the double of the coefficient of the squares
The fourth is the Characteristick 6

3. Of the biquadrates

[...etc.]

[100] CUL Add MS 9597/9/15, f. 15.

These examples and rules fill pages 15 to 17 of Mercator's treatise, and all are from (W4). Page 18 is crossed out. Page 19 is then headed 'Problema Arithmetica ad doctrinam de differentium Progressionibus pertinens', and this and page 20 contain an exact copy of (W1). In fact the original, (W1) itself, is neatly folded into the back of the treatise, presumably placed there by Collins himself.

The next ten pages of the treatise, 21 to 30, contain the material from (W2) but in a different order: Mercator's pages 21 to 26 are a copy of (W2b), in a slightly changed order with the theorems now placed before the problems and worked examples, and pages 27 to 30 are a copy of the four problems from (W2a).

These pages are followed in Mercator's treatise by yet more material from Warner: Mercator's page 31 is headed 'Methodus Calculatoria qua Canon analogicus ...' and is an exact copy of (W3), Warner's two examples of interpolating a table of antilogarithms. The remaining seven pages of Mercator's treatise (32 to 38) are then devoted to discussion of antilogarithms, with a short table of antilogarithms at the bottom of page 38.

Thus 16 pages of Mercator's 'Doctrine of differences' are copied from Warner's manuscripts and include all of items (W1) to (W4), with only slight changes in the internal ordering. This discovery has been fortuitous: Mercator's treatise came to light only after it was acquired by Cambridge University in 2000, and before that no link between Warner and Mercator was suspected. Even now the relevant manuscripts are scattered through different volumes in two separate libraries, making direct comparisons impossible. Of course this finding raises the new question of how and when Mercator, who arrived in England ten years after Warner's death, came to know his work so well?

The answer can only be from Pell or Collins, or perhaps both. Mercator and Pell probably met in the Netherlands in 1644, because Pell told Collins in 1666 that he had known Mercator for 22 years.[101] They would certainly have become reacquainted after Pell's return to England from Switzerland in 1658, since both were eventually members of the Royal Society. However, Pell's relationship with Mercator seems to have been somewhat prickly. Pell described the contents of Mercator's *Logarithmotechnia*, for instance, dismissively as 'crudities'.[102] Pell may have held on to some isolated sheets from Warner's manuscripts, but after 1650 most of Warner's surviving papers were held by Thorndike. In 1667 Thorndike in turn lent several bundles of them to Collins,[103] who therefore seems to be the more likely source of Mercator's familiarity with them.

Early in 1668 Collins wrote as follows to James Gregory:[104]

> I have some papers of Mr Warner deceased, wherein he proves if parallels be drawn to an asymptote, so as to divide the other into equal parts, the spaces between them, the hyperbola, and asymptote, are in musical progression ...

The second part of (W4) appears, out of place, amidst what was otherwise Warner's

[101] BL Add MS 4278, f. 118v.

[102] BL Add MS 4415, f. 2.

[103] For 'An inventorie of the papers of Mr Warner' see BL Add MS 4394, f. 106.

[104] Gregory 1939, 45.

treatment of the hyperbola, and so if Collins knew about the hyperbola he must also have seen (W4). We know that during 1668 or early 1669 Collins became interested in solving equations by 'progressional differences', because he referred to the method in his notes on 'Improvements of Algebra in England', where he presumed that it was the purpose of the lengthy tables created by Pell. At the end of those notes he gave a few simple examples of constant difference tables for polynomials, and they are precisely those that appear both in Warner's (W4) and in Mercator's 'Doctrine of differences', namely for the three equations $aa + 3a = 130$, $aaa + 2aa + 3a = 1230$, and $aaaa + 1aaa + 3aa + 2a =$ 'a number'. Collins also quoted Warner's rules for constructing the first row of each table.[105]

Collins also mentioned the method briefly in the paper he published in the *Philosophical Transactions* in April 1669, where he gave a table of first, second, and third differences for the equation $aaa - 3aa + 4a = N$. Then in June the following year he sent his 'Narrative about equations' to Isaac Barrow.[106] Although he had repeatedly complained that he could not extract the method of differences from Pell, in the first part of the 'Narrative' he was nevertheless able to give the example shown in Table 36,

+40			
	+16		
+56		−18	
	−2		+6
+54		−12	
	−14		+6
+40		−6	
	−20		+6
+20		∓0	
	−20		+6
±0		+6	
	−14		+6
−14		+12	
	−2		+6
−16		+18	
	+16		+6
∓0		+24	
	+40		+6
+40		+30	

Table 36. Values of $a^3 - 15a^2 + 54a$ for integer values of a from 1 to 17, together with first, second, and third differences.

[105] Compare Mercator, CUL Add MS 9597/9/15, ff. 15–17, and Collins, BL Add MS 4474, f. 3v.

[106] Collins, 'Narrative about Aequations', 1670, in Gregory 1939, 116–117; see also Rigaud 1841, II, 601–603; there is also a manuscript copy in Aubrey's hand in Worcester College, Oxford, MS 64. For the full 'Narrative' see Gregory 1939, 113–118 and 142–145.

for solving the equation $a^3 - 15a^2 + 54a = N$. The column on the left contains values of $a^3 - 15a^2 + 54a$ for $a = 1$ to $a = 17$, followed by first, second, and third differences.

In the second part of the 'Narrative', which is headed 'Concerning rankes of numbers whose last differences are equal', Collins was able to go further, and stated that the method of interpolation could be carried out 'by ayd of a table of figurate Numbers'. More precisely,[107]

> out of the given differences, to raise the heads of the new differences that without any correction (which Briggs his method requires) shall carry on the work.

In Harriot's treatment, 'raising the heads of the new differences out of the given differences', was precisely the problem of finding new column headings $b, c, d, \ldots$ from the given differences $B, C, D, \ldots$. Collins did not try to explain how this should be done, but he does at least seem to have understood the problem, and was able to compare what was essentially Harriot's method with the equivalent method given by Briggs.

Could it be that Collins had finally learned from Warner's papers and with help from Mercator what he had never been able to learn from Pell? Mercator had already introduced some simple difference tables into his *Logarithmotechnia*, published in 1668 but written in 1667; there he used them in the calculations of logarithms.[108] Thus by 1668, or early 1669 at the latest, Mercator and Collins were both interested in difference methods, Mercator in connection with the calculation of logarithms and Collins for solving equations. It was probably this common interest that led Collins to show Mercator what he had found in Warner's papers, and it was probably also Collins who persuaded Mercator to write his 'Doctrine of differences' to summarize that material. We may therefore tentatively date Mercator's 'Doctrine of differences' to sometime between 1668 and 1670.

Through his papers, Warner passed on the method of differences not only to Pell, whom he knew personally, but indirectly and long after his death to Collins and Mercator also. Warner in turn had learned almost everything he knew from Harriot. The ideas and notation of Harriot's 'Magisteria', often adapted and simplified, and sometimes garbled, are nonetheless still clearly visible in Mercator's 'Doctrine of differences'. Mercator's treatise is thus of unique historical interest as evidence that Harriot's method of differences was still in use amongst a small group of English mathematicians almost sixty years after Harriot invented it.

Isaac Newton (1642–1727)

In the early 1650s, as part of his attempt to find the quadrature of the circle, John Wallis found that he needed to interpolate between the rows and columns of a table of generalized triangular numbers, which he laid out as in Table 37 (see also Table 1).[109]

---

[107] Gregory 1939, 144–145.
[108] Mercator 1668, 11–14 and 23–24.
[109] Wallis 1656, Proposition 169.

1	1	1	1	1	1	...
1	2	3	4	5	6	...
1	3	6	10	15	21	...
1	4	10	20	35	56	...
1	5	15	35	70	126	...
1	6	21	56	126	252	...
.	.	.	.	.	.	
.	.	.	.	.	.	

Table 37. Wallis's table of triangular numbers.

With considerable toil, Wallis rediscovered Harriot's formulae for figurate numbers,[110] and then used them to interpolate 'halfway' values into the rows as shown in Table 38, the first time the general triangular numbers had been treated as anything other than integers.[111]

1	1	1	1	1	1	1	1	1	1	...	1
$\frac{1}{2}$	1	$1\frac{1}{2}$	2	$2\frac{1}{2}$	3	$3\frac{1}{2}$	4	$4\frac{1}{2}$	5	...	l
$\frac{3}{8}$	1	$1\frac{7}{8}$	3	$4\frac{3}{8}$	6	$7\frac{7}{8}$	10	$12\frac{3}{8}$	15	...	$\frac{l^2+l}{2}$
$\frac{15}{48}$	1	$2\frac{9}{48}$	4	$6\frac{27}{48}$	10	$14\frac{21}{48}$	20	$26\frac{39}{48}$	35	...	$\frac{l^3+3l^2+2l}{6}$
$\frac{105}{384}$	1	$2\frac{177}{384}$	5	$9\frac{9}{384}$	15	$23\frac{177}{384}$	35	$50\frac{105}{384}$	70	...	$\frac{l^4+6l^3+11l^2+6l}{24}$
.	.	.	.	.	.	.	.	.	.	...	

Table 38. Wallis's table of triangular numbers, interpolated using formulae.

Wallis next experimented with writing each number as a multiple of previous ones, and observed that the third row, for example, is generated by successive stages of the multiplication $1 \times \frac{3}{2} \times \frac{5}{4} \times \frac{7}{6} \times \frac{8}{7} \times \dots$, with similar patterns for the other rows.[112]

In 1656 Wallis published his findings in the *Arithmetica infinitorum*, and just eight years later his book was studied by Isaac Newton, then twenty-two years old. Newton, following a similar line of thought, relied at first on Wallis's interpolated entries, but by the autumn of 1665 he had seen a much more efficient way of proceeding. Amongst his manuscripts we find the following table. Newton had started from the binomial coefficients in the expansion of $(1 + x)^n$, for n an integer, which appear listed in the odd-numbered columns. Next, like Wallis, he interpolated additional entries in the even-numbered columns, as shown in Table 39.[113]

[110] Wallis 1656, Propositions 171–182.

[111] Wallis 1656, Proposition 184. Wallis also interpolated into the columns, so that his table retained its symmetry, but here we restrict ourselves to row interpolation only.

[112] Wallis 1656, Proposition 188.

1	1	1	1	1	1	1	1	1	1	1
1	$\frac{3}{2}$	2	$\frac{5}{2}$	3	$\frac{7}{2}$	4	$\frac{9}{2}$	5	$\frac{11}{2}$	6
0	$\frac{3}{8}$	1	$\frac{15}{8}$	3	$\frac{35}{8}$	6	$\frac{63}{8}$	10	$\frac{99}{8}$	15
0	$-\frac{1}{16}$	0	$\frac{5}{16}$	1	$\frac{35}{16}$	4	$\frac{105}{16}$	10	$\frac{231}{16}$	20
0	$\frac{3}{128}$	0	$-\frac{5}{128}$	0	$\frac{35}{128}$	1	$\frac{315}{128}$	5	$\frac{1155}{128}$	15
0	$-\frac{3}{256}$	0	$\frac{3}{256}$	0	$-\frac{7}{256}$	0	$\frac{63}{256}$	1	$\frac{693}{256}$	6
0	$\frac{7}{1024}$	0	$-\frac{5}{1024}$	0	$\frac{7}{1024}$	0	$-\frac{21}{1024}$	0	$\frac{231}{1024}$	1

Table 39. Newton's table of binomial coefficients, interpolated.

The entries above and including the diagonal line of 1s are those calculated by Wallis (in Table 38) but in an altered layout. The entries below the diagonal, including several 0s, have been calculated from a rule and a chart (Table 40) that Newton wrote alongside:

> The property of which table is yt ye sume of any figure and ye figure above it is equall to ye figure next after it save one.

Then he added:

> Also ye numerall progressions [that is, the rows of the table] are of these forms.

a	a	a	a
b	$a+b$	$2a+b$	$3a+b$
c	$b+c$	$a+2b+c$	$3a+3b+c$
d	$c+d$	$b+2c+d$	$a+3b+3c+d$
e	$d+e$	$c+2d+e$	$b+3c+3d+e$

Table 40. Newton's algebraic difference table.

Table 40 is, of course, just like those written by Harriot (see Table 4), but with the constant difference a now running horizontally across the top.

Newton now recognized a new method of interpolation even simpler than Harriot's, and very much superior to Wallis's. Taking, for example, the sequence of triangular numbers 0, 1, 3, 6, 10, Newton re-wrote it with spaces between, as

$$0 \quad \star \quad 1 \quad \star \quad 3 \quad \star \quad 6 \quad \star \quad 10 \quad \ldots$$

where each $\star$ is to be replaced by some interpolated value. Equating this with the third row of his difference table (Table 40), he obtained the equations

[113] CUL Add MS 3958.3:72, reproduced in Newton 1967–81, I, frontispiece. In the original there are four further columns to the left, with which we need not be concerned here.

$$\begin{aligned} 0 &= c \\ \star &= b + c \\ 1 &= a + 2b + c \\ \star &= 3a + 3b + c \\ 3 &= 6a + 4b + c \\ \star &= 10a + 5b + c \\ 6 &= 15a + 6b + c. \end{aligned}$$

These are easily solved to give $c = 0$, $b = \frac{3}{8}$, $a = \frac{1}{4}$, and hence every other value in the row.

The method was so simple that Newton was able to interpolate not just one but any number of new values between each integer entry. The integer columns, as we have already observed, contain binomial coefficients for the expansion of $(1 + x)^n$, for n a positive integer, and Newton assumed that the interpolated entries gave him the corresponding coefficients when n is a fraction. Thus he was led to the formula

$$\frac{n(n-1)(n-2)\dots(n-r+1)}{1.2.3.\dots.r}$$

for the coefficient of x^r in the expansion of $(1 + x)^n$ for *all* values of n, an empirical discovery of the general binomial theorem, and a step of enormous importance in seventeenth-century mathematics.

Newton's difference table is so similar to those written down by Harriot and his successors, that one is bound to ask whether he had any idea of those earlier investigations. The answer would seem to be almost certainly that he did not. As a young student, working largely in isolation in Lincolnshire, he would not have seen or even heard of the unpublished manuscripts of Torporley, Warner, or Pell. He may have seen Briggs' *Arithmetica logarithmica* but he did not mention Briggs as one of the authors he had read in his early years,[114] and in any case Briggs gave only numerical examples, not general algebraic tables. Mercator's tables, the first to appear in print, were not published until four years after Newton had written out his own. It would seem then, that Newton, like Harriot, moved rapidly and independently to a general algebraic difference table.

In 1676 Newton returned to interpolation by finite differences for calculating tables of roots (just as Harriot had done when calculating meridional parts) and described the process of inserting 99 interpolation points by the wonderful name of 'subcenturying'.[115] A hint of the method appeared in print a few years later in the *Principia*,[116] where Newton showed how to determine paths of comets from a small number of observations. During the 1690s he developed this idea further, moving on to a more general theory of curve-fitting; his tables and formulae throughout this work are strikingly similar to Harriot's.[117] Further work completed in 1710 was published by William Jones the following year.[118]

[114]Newton 1967–81, IV, 3–13.

[115]Newton 1967–81, IV, 14–73; see especially 52–53 for a splendid example of a difference table.

[116]Newton 1687, Book III, Lemma V.

[117]Newton 1967–81, VI, 672–681.

James Gregory (1638–75)

By the end of 1670 James Gregory too had arrived at a form of what is now called the Newton–Gregory interpolation formula, but unlike Newton who left plenty of manuscript evidence, Gregory left little in the way of explanation. The formula itself was first given in a letter to Collins in 1670.[119] In it Gregory denoted the abscissa and ordinate at a given point by c and d respectively, and the abscissa at a second point by a (with the assumption that $a > c$). To find the ordinate at a he denoted 'the first of the first differences = f, of the second differences = h, of the third differences = i', and so on.[120] He also defined a sequence

$$\begin{aligned}
&\frac{a}{c} \\
&\frac{b}{c} = \frac{a(a-c)}{c.2c} \\
&\frac{k}{c} = \frac{a(a-c)(a-2c)}{c.2c.3c} \\
&\frac{l}{c} = \frac{a(a-c)(a-2c)(a-3c)}{c.2c.3c.4c} \\
&\ldots
\end{aligned}$$

and then stated that the ordinate at a is given by

$$\frac{ad}{c} + \frac{bf}{c} + \frac{kh}{c} + \frac{li}{c} + \text{etc.}$$

Expanding the individual terms, this appears in more familiar form as:

$$\frac{ad}{c} + \frac{a(a-c)f}{c.2c} + \frac{a(a-c)(a-2c)h}{c.2c.3c} + \frac{a(a-c)(a-2c)(a-3c)i}{c.2c.3c.4c} + \cdots$$

Presumably Gregory obtained these formulae in much the same way as Brook Taylor in his *Methodus incrementorum* forty-five years later, by summing successive differences.[121] Taylor in the early eighteenth century described the formulae for binomial coefficients as the 'Newtonian theorem', but Gregory in 1670 did not yet know of Newton's work, and worked out the formulae for himself, as Harriot had done half a century earlier. The most striking difference between Gregory's work and Harriot's, of course, is that Gregory was now working with infinite series and so no longer needed to

[118]Newton 1967–81, VIII, 236–257; Newton 1711, 93–101.

[119]Gregory to Collins, 23 November 1670, in Gregory 1939, 119–120.

[120]'primam ex differentiis primis = f, ex differentiis secundis = h, ex differentiis tertiis = i, ex differentiis quartis = k, ex differentiis quintis = l et omnes differentias affici signo +'; Gregory 1939, 119.

[121]Taylor 1715, 21.

assume, as Harriot had done, that successive differences were eventually constant.[122] He sent Collins two enclosures demonstrating the application of his method to series for sines and logarithms, and later letters carried further results on infinite series.[123]

Conclusion

By the mid 1670s European mathematics was advancing rapidly, and the interlinked discoveries of calculus and infinite series in the 1660s and 1670s transformed the subject almost beyond recognition. Harriot's work now appears as no more than a small stream compared with the flood of ideas that came later. It cannot be claimed that the 'Magisteria' influenced Newton or Gregory, who developed their ideas powerfully and independently and with all the advantages of a further half century of mathematics to draw upon. Nevertheless, the clarity of Harriot's exposition ensured the survival of his interpolation method for nearly sixty years, during which time Torporley, possibly Briggs, Warner, Pell, Collins, and Mercator all continued to work with it. Because most of their work on the subject was never published, and because their manuscripts are now scattered between at least four different libraries, the long running interest in constant difference methods in England up to 1670, and the ways in which those methods were passed from one mathematician to another, have until now gone unrecognized. Inevitably many details of the story are now lost, but we can say without any doubt that by 1611 or soon afterwards Harriot had worked out the essential ideas of an interpolation theory that remained in use for half a century or more, and not until the arrival of Newton and Gregory was it finally surpassed.

[122] There is just one example in Harriot's manuscripts of a similar derivation of an infinite series, in BL Add MS 6782, ff. 67–72.

[123] Gregory to Collins, 23 November 1670, in Gregory 1939, 127–133.

Bibliography

Manuscripts

British Library: Additional MSS 4394–4396 (Warner); 4365, 4413, 4424, 4431 (Pell); 4474 (Collins); 6782–6789 (Harriot); and Harleian MS 6083 (Cavendish). Copies of Add MSS 6782–6789 and Harleian MS 6083 are held in the Bodleian Library, Oxford, in the Modern Papers Reading Room in boxes Dep.b.178–183, and in the University of Delaware Library, Special Collections, John Shirley Papers, Boxes 11–26, 28–29.

Cambridge University Library: Macclesfield Papers, CUL Additional MSS 9597/17/28 (Torporley); 9597/9/15 (Collins, Mercator, Warner); and Additional MS 3958 (Newton).

Lambeth Palace Library: Sion College MS Arc L.40.2/L.40 (Torporley).

Petworth House, Sussex: MSS 240 and 241 (Harriot). Copies of the Petworth papers are held in the West Sussex Record Office, Chichester, and also in the University of Delaware Library, Special Collections, John Shirley Papers, Boxes 31–33.

Sheffield University Library (and on CD): *Ephemerides* 29/3 (Hartlib).

Books and articles

Beery, Janet, 'Navigating between triangular numbers and trigonometric tables: how Thomas Harriot developed his interpolation formulas', *Proceedings of the Canadian Society for the History and Philosophy of Mathematics: thirty-second Annual Meeting*, 2007, 37–47.

Boethius, Anicius Manlius Severinus, *De institutione arithmeticae*, translated by Michael Masi as *Boethian number theory: a translation of the* De institutione arithmetica, Studies in Classical Antiquity, vol. 6, Amsterdam: Rodopi, 1983.

Briggs, Henry, *Logarithmorum chilias prima*, London, 1617.

Briggs, Henry, *Arithmetica logarithmica*, London, 1624.

Briggs, Henry, *Trigonometria britannica*, completed by Henry Gellibrand, Gouda, 1633.

Cardano, Girolamo, *Opus novum de proportionibus numerorum* ..., Basel, 1570.

Collins, John, 'An account concerning the resolution of equations in numbers', *Philosophical Transactions*, 4 (1669), 929–934.

Edwards, A. W. F., *Pascal's arithmetical triangle*, London: C. Griffin and New York: Oxford University Press, 1987.

Etaples, Jacques Lefevre, *Epitome, compendiosaque introductio in libros arithmeticos diui Seuerini Boetij: adiecto familiari commentario dilucidata*, Paris, 1503; 1522.

Gregory, James, *James Gregory: tercentenary memorial volume*, edited by H. W. Turnbull, London, 1939.

Hakewill, George, *An apologie of the power and providence of God in the government of the world*, second edition, Oxford, 1630.

Halliwell-Phillips, James Orchard (ed), *A collection of letters illustrative of the progress of science in England: from the reign of Queen Elizabeth to that of Charles the Second*, London: R. and J. E. Taylor, 1841.

Hofmann, Joseph E, *Leibniz in Paris 1672–1676: his growth to mathematical maturity*, Cambridge: Cambridge University Press, 1974.

Kepler, Johann, *Joannis Kepleri astronomi opera omnia*, edited by Christian Frisch, 8 vols., Frankfurt, 1858–72.

Malcolm, Noel, and Stedall, Jacqueline, *John Pell (1611–1685) and his correspondence with Sir Charles Cavendish: the mental world of an early modern mathematician*, Oxford: Oxford University Press, 2005.

Maurolico, Franciscus, *Arithmeticorum libri duo*, Venice, 1575.

Mercator, Nicolaus, *Logarithmotechnia*, London, 1668.

Mouton, Gabriel, *Observationes diametrorum solis et lunae apparentium*, Lyon, 1670.

Newton, Isaac, *Philosophiae naturalis principia mathematica*, London, 1687.

Newton, Isaac, *Analysis per quantitatum series, fluxiones, ac differentias*, edited by W. Jones, London, 1711.

Newton, Isaac, *The mathematical papers of Isaac Newton*, edited by D. T. Whiteside, 8 vols., Cambridge: Cambridge University Press, 1967–1981.

Pepper, Jon V., 'Harriot's unpublished papers', *History of Science*, 6 (1967), 17–40.

Pepper, Jon V., 'Harriot's calculations of the meridional parts as logarithmic tangents', *Archive for History of Exact Sciences*, 4 (1968), 359–413.

Rigaud, Stephen Peter (ed), *Correspondence of scientific men of the seventeenth century*, Oxford, 1841.

Shirley, John W., *Thomas Harriot: a biography*, Oxford: Clarendon Press, 1983.

Stapulensis, Jacob Faber, see Etaples, Jacques Lefevre.

Stedall, Jacqueline, *The greate invention of algebra: Thomas Harriot's treatise on equations*, Oxford: Oxford University Press, 2003.

Stevin, Simon, *L'arithmetique: aussi l'algebre. Ensemble les quatre premiers liures d'Algebre de Diophante d'Alexandrie*, Leiden, 1585.

Stifel, Michael, *Arithmetica integra*, Nuremberg, 1544.

Tanner, Rosalind Cecilia H, 'Nathaniel Torporley's "Congestor analyticus" and Thomas Harriot's "De triangulis laterum rationalium"', *Annals of Science*, 34 (1977), 393–428.

Taylor, Brook, *Methodus incrementorum*, London, 1715.

Viète, François, *Opera mathematica*, Leiden, 1646.

Wallis, John, *Arithmetica infinitorum*, in *Operum mathematicorum pars altera*, Oxford, 1656.

Wallis, John, *Opera mathematica*, 3 vols., Oxford, 1693–99.

Title page [Folio 107; half sheet]*

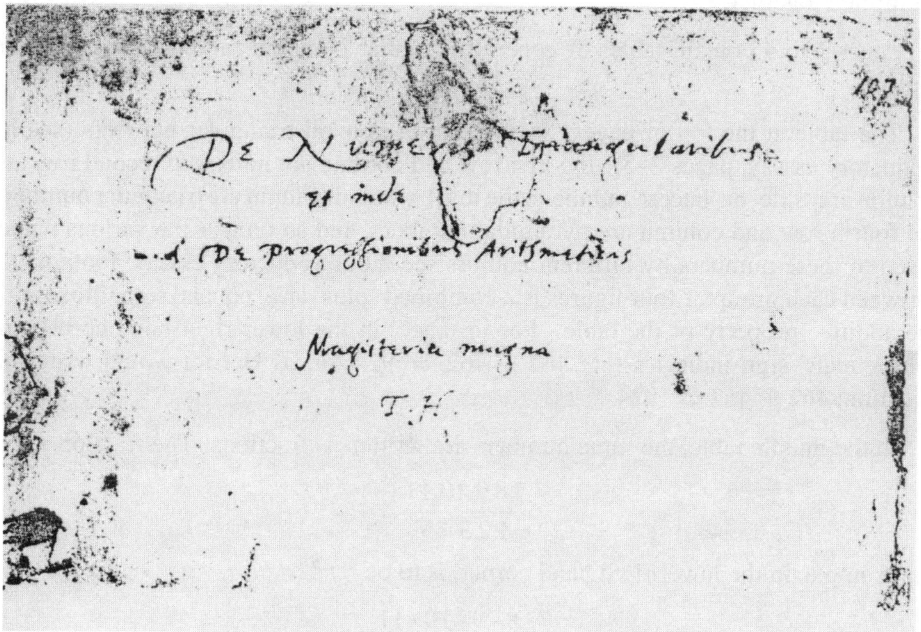

De Numeris Triangularibus
Et inde
De Progressionibus Arithmeticis
Magisteria magna
T. H.

On Triangular Numbers
And thence
On Arithmetic Progressions
The great Doctrine
T[homas] H[arriot]

*The following pages from BL Add MS 6782 are reproduced by kind permission of The British Library.

Page 1 [Folio 108]

Notation: Harriot always used his own equality sign $\overline{\underline{\|}}$ rather than Recorde's $=$. Throughout the treatise n is to be read as a positive integer. Terms stacked vertically (sometimes with a vertical line down the right hand side) are to be understood to be multiplied together.

Pages 1 to 4 contain tables of general triangular numbers and corresponding formulae.

The table at the top of page 1 is an array of general triangular numbers (see introductory esssay, pages 2–5): the first row and column are units, the second row and column are 'side' or 'lateral' numbers, the third row and column are triangular numbers, the fourth row and column are pyramidal numbers, and so on (for the various names given to these numbers by different authors see the introductory esssay, footnote 4). Between each group of four figures is a combined 'plus' and 'equals' sign, illustrating the additive property of the table. For instance, in the lower right hand corner, the 'plus/equals' sign indicates that $462 + 462 = 924$, or, as Harriot would write this equation, $462 + 462 \overline{\underline{\|}} 924$.

In the middle table, the same numbers are written as fractions. The fraction

$$\frac{7\,8\,9{,}10{,}11}{1\,2\,3\,4\,5},$$

for example, in the lower right hand corner, is to be read as

$$\frac{7 \cdot 8 \cdot 9 \cdot 10 \cdot 11}{1 \cdot 2 \cdot 3 \cdot 4 \cdot 5}.$$

This table therefore displays the pattern Harriot would use to write a general formula for each entry. In the lower table Harriot wrote general algebraic formulae for the numbers in the nth row. The fourth number in the nth row, for example, is

$$\frac{n(n+1)(n+2)}{1 \cdot 2 \cdot 3}.$$

Harriot's '&c' ('etc.') here and throughout the treatise indicates that the tables can be extended indefinitely in the directions specified.

Related material: Similar tables appear elsewhere in Harriot's papers at British Library Add MSS 6782, f. 219; 6784, ff. 207v and 210; 6785, ff. 4–4v and 83–83v; and 6786, f. 417.

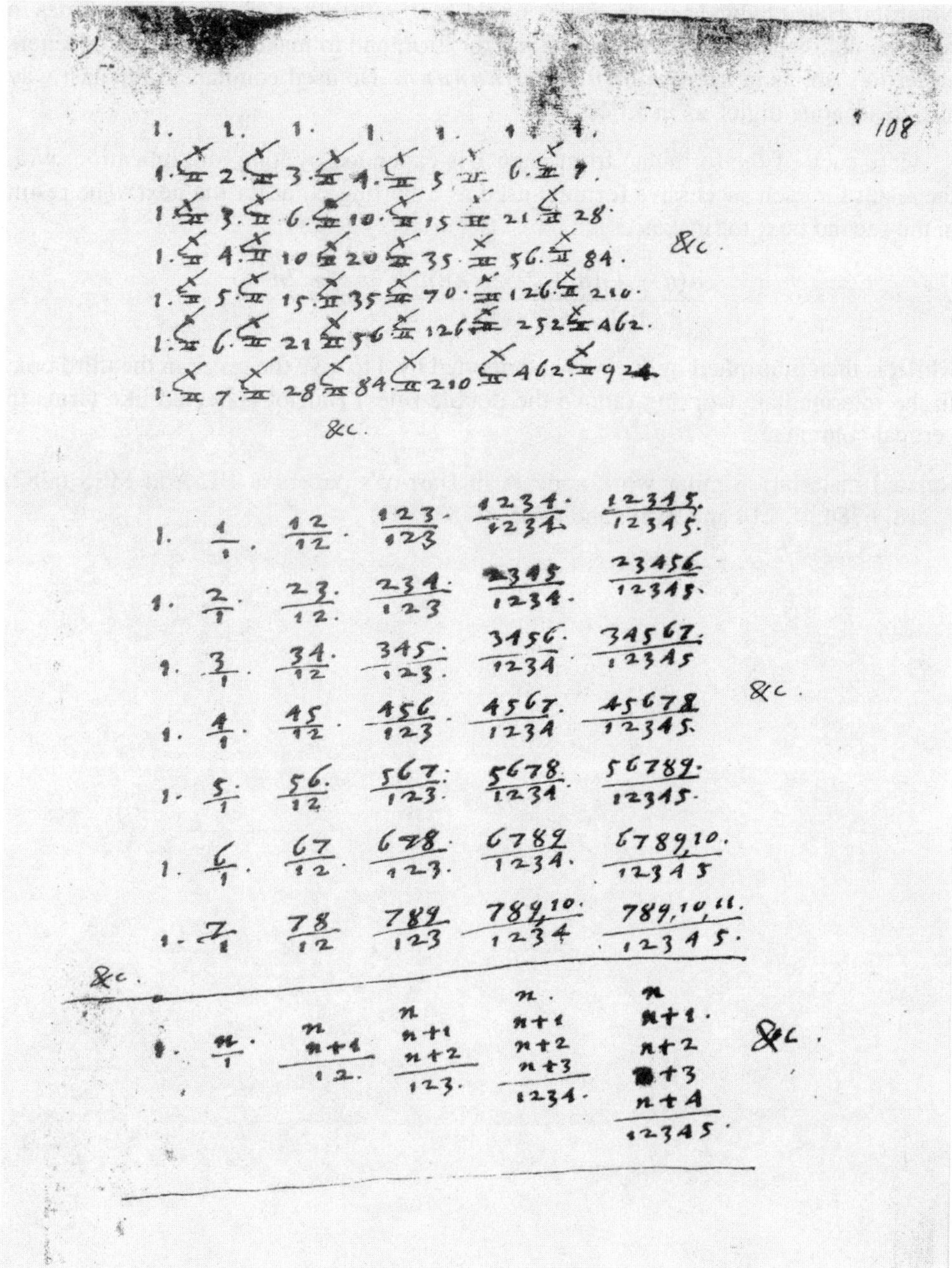

Page 2 [Folio 109]

Notation: Harriot always wrote, for example, nnn rather than n^3. He used commas to separate numbers from letters, as in 2,nn for 2nn, and to make long strings of letters easier to read, as in nn,nnn,nnn for $nnnnnnnn$. He used commas in this last way also to separate digits, as in 5,040.

Here each of the formulae from page 1 is expanded by long multiplication, with the result for each successive formula used as a starting point for the next. The result in the second box, for instance, is

$$\frac{n(n+1)(n+2)}{1\cdot 2\cdot 3} = \frac{nnn + 3nn + 2n}{6},$$

which is then multiplied by $(n+3)$ and divided by 4 to give the result in the third box. In the intermediate working (above the double rules) Harriot collected like terms in vertical columns.

Related material: Similar work appears in Harriot's papers at BL Add MSS 6782, f. 218; 6784, ff. 214 and 215v; and 6785, ff. 83–83v.

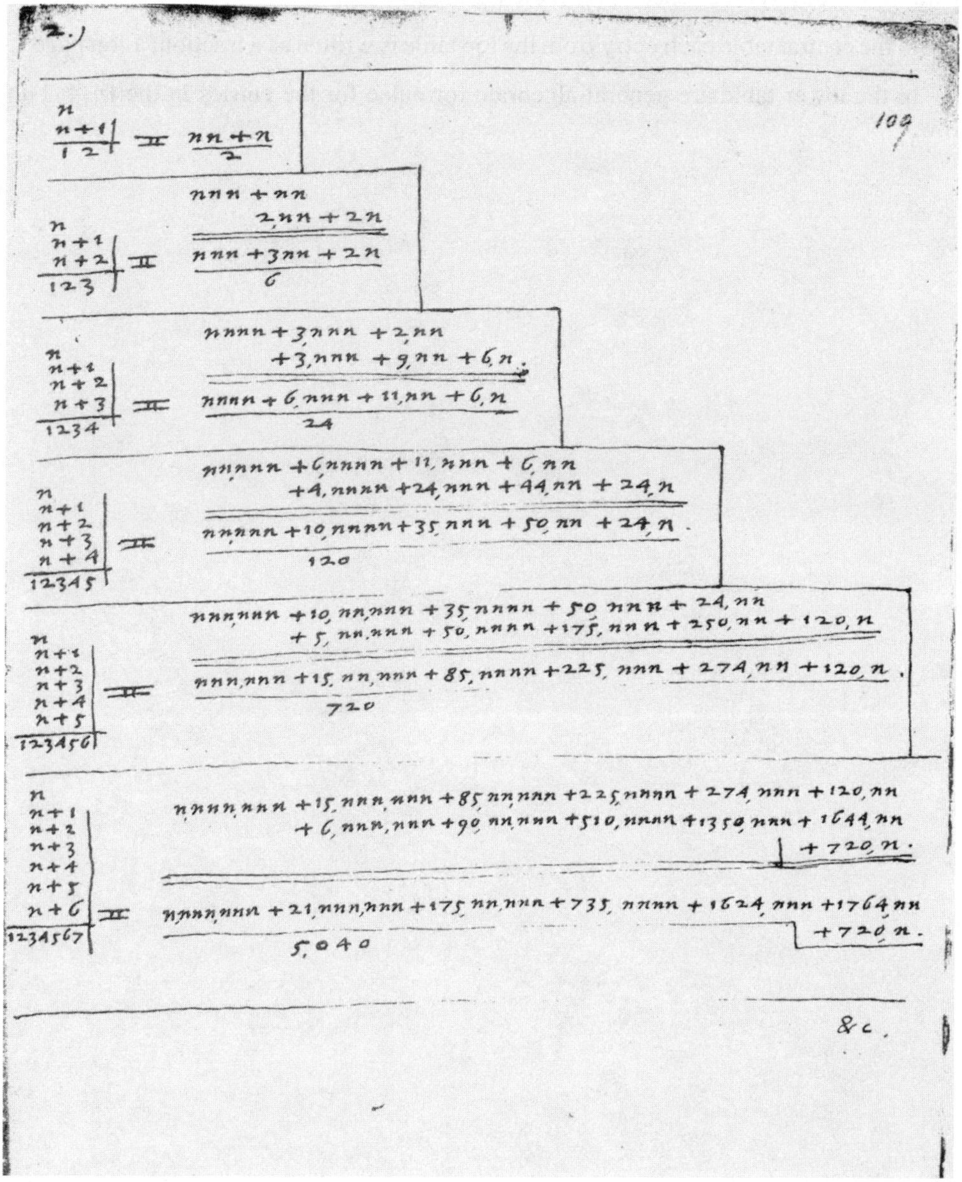

2.) 109

n, n+1 / 12 = (nn+n)/2

n, n+1, n+2 / 123 = nnn + nn; 2,nn + 2n; (nnn + 3nn + 2n)/6

n, n+1, n+2, n+3 / 1234 = nnnn + 3,nnn + 2,nn; + 3,nnn + 9,nn + 6,n; (nnnn + 6,nnn + 11,nn + 6,n)/24

n, n+1, n+2, n+3, n+4 / 12345 = nnnnn + 6,nnnn + 11,nnn + 6,nn; + 4,nnnn + 24,nnn + 44,nn + 24,n; (nnnnn + 10,nnnn + 35,nnn + 50,nn + 24,n.)/120

n, n+1, n+2, n+3, n+4, n+5 / 123456 = nnn,nnn + 10,nn,nnn + 35,nnnn + 50,nnn + 24,nn; + 5,nn,nnn + 50,nnnn + 175,nnn + 250,nn + 120,n; (nnn,nnn + 15,nn,nnn + 85,nnnn + 225,nnn + 274,nn + 120,n.)/720

n, n+1, n+2, n+3, n+4, n+5, n+6 / 1234567 = n,nnn,nnn + 15,nnn,nnn + 85,nn,nnn + 225,nnnn + 274,nnn + 120,nn; + 6,nnn,nnn + 90,nn,nnn + 510,nnnn + 1350,nnn + 1644,nn + 720,n.; (nnnn,nnn + 21,nnn,nnn + 175,nn,nnn + 735,nnnn + 1624,nnn + 1764,nn + 720,n.)/5,040

&c.

Page 3 [Folio 110]

In the top table the triangular numbers from page 1 are rearranged into a triangular pattern, with the sum of each row on the right.

In the central table each entry from the top table is written as a fraction, as on page 1.

In the lower table are general algebraic formulae for the entries in the $(n+1)$th row.

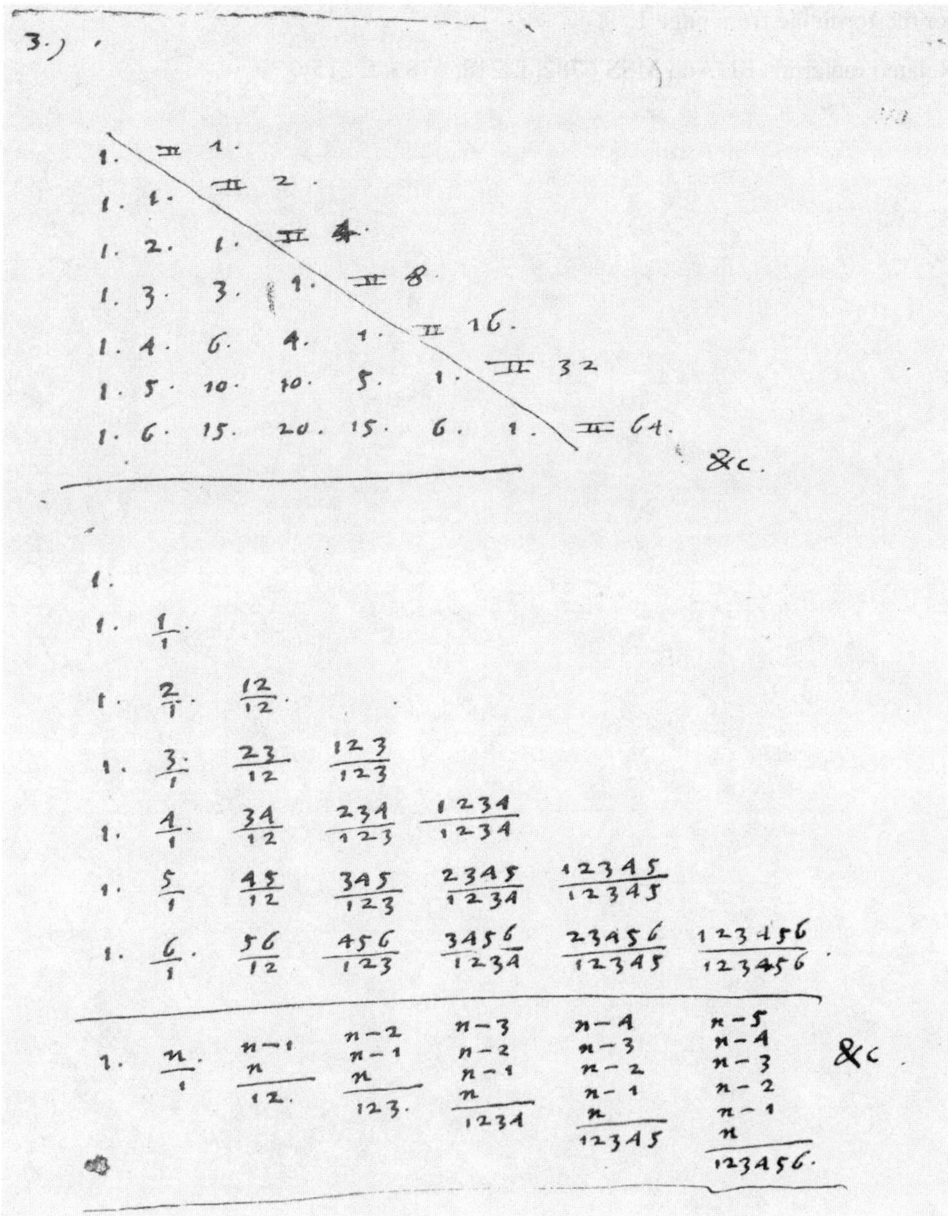

3.)

1. = 1.
1. 1. = 2.
1. 2. 1. = 4.
1. 3. 3. 1. = 8.
1. 4. 6. 4. 1. = 16.
1. 5. 10. 10. 5. 1. = 32
1. 6. 15. 20. 15. 6. 1. = 64.
&c.

1.
1. $\frac{1}{1}$.
1. $\frac{2}{1}$. $\frac{12}{12}$.
1. $\frac{3}{1}$ $\frac{23}{12}$ $\frac{123}{123}$
1. $\frac{4}{1}$. $\frac{34}{12}$ $\frac{234}{123}$ $\frac{1234}{1234}$
1. $\frac{5}{1}$ $\frac{45}{12}$ $\frac{345}{123}$ $\frac{2345}{1234}$ $\frac{12345}{12345}$
1. $\frac{6}{1}$. $\frac{56}{12}$ $\frac{456}{123}$ $\frac{3456}{1234}$ $\frac{23456}{12345}$ $\frac{123456}{123456}$.

1. $\frac{n}{1}$. $\frac{n-1 \quad n}{12.}$ $\frac{n-2 \quad n-1 \quad n}{123.}$ $\frac{n-3 \quad n-2 \quad n-1 \quad n}{1234}$ $\frac{n-4 \quad n-3 \quad n-2 \quad n-1 \quad n}{12345}$ $\frac{n-5 \quad n-4 \quad n-3 \quad n-2 \quad n-1 \quad n}{123456.}$ &c.

Page 4 [Folio 111]

Harriot has expanded his formulae from the bottom of page 3, just as he did on page 2 for the formulae from page 1.

Related material: BL Add MSS 6782, f. 218; 6784, f. 215v.

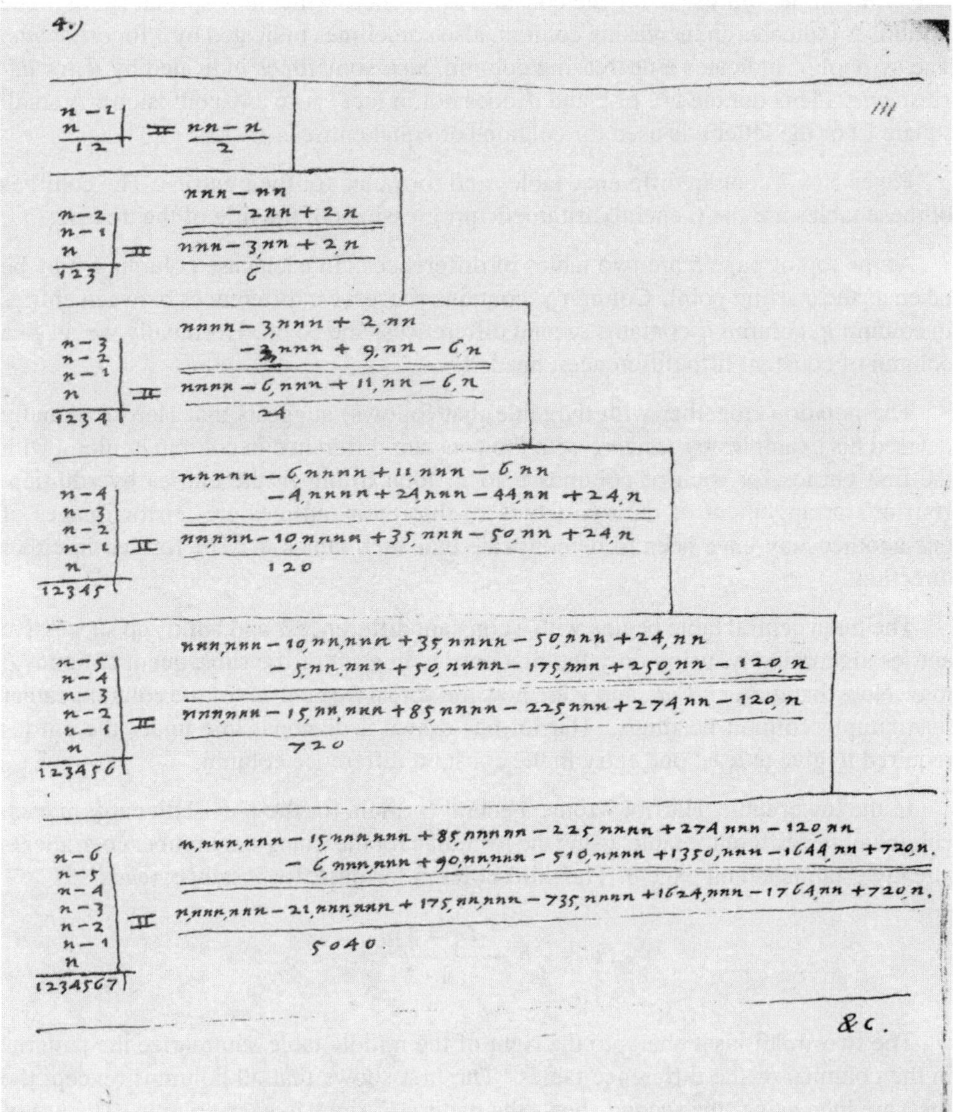

4.)

III

$$\begin{array}{c|} n-1\\ n\\ \hline 1\,2 \end{array} \;\sqcap\; \frac{nn-n}{2}$$

$$\begin{array}{c|} n-2\\ n-1\\ n\\ \hline 1\,2\,3 \end{array} \;\sqcap\; \begin{array}{l} nnn-nn\\ \quad -2nn+2n\\ \hline \dfrac{nnn-3nn+2n}{6} \end{array}$$

$$\begin{array}{c|} n-3\\ n-2\\ n-1\\ n\\ \hline 1\,2\,3\,4 \end{array} \;\sqcap\; \begin{array}{l} nnnn-3nnn+2nn\\ \quad -3nnn+9nn-6n\\ \hline \dfrac{nnnn-6nnn+11nn-6n}{24} \end{array}$$

$$\begin{array}{c|} n-4\\ n-3\\ n-2\\ n-1\\ n\\ \hline 1\,2\,3\,4\,5 \end{array} \;\sqcap\; \begin{array}{l} nnnnn-6nnnn+11nnn-6nn\\ \quad -4nnnn+24nnn-44nn+24n\\ \hline \dfrac{nnnnn-10nnnn+35nnn-50nn+24n}{120} \end{array}$$

$$\begin{array}{c|} n-5\\ n-4\\ n-3\\ n-2\\ n-1\\ n\\ \hline 1\,2\,3\,4\,5\,6 \end{array} \;\sqcap\; \begin{array}{l} nnnnnn-10nnnnn+35nnnn-50nnn+24nn\\ \quad -5nnnnn+50nnnn-175nnn+250nn-120n\\ \hline \dfrac{nnnnnn-15nnnnn+85nnnn-225nnn+274nn-120n}{720} \end{array}$$

$$\begin{array}{c|} n-6\\ n-5\\ n-4\\ n-3\\ n-2\\ n-1\\ n\\ \hline 1\,2\,3\,4\,5\,6\,7 \end{array} \;\sqcap\; \begin{array}{l} nnnnnnn-15nnnnnn+85nnnnn-225nnnn+274nnn-120nn\\ \quad -6nnnnnn+90nnnnn-510nnnn+1350nnn-1644nn+720n.\\ \hline \dfrac{nnnnnnn-21nnnnnn+175nnnnn-735nnnn+1624nnn-1764nn+720n.}{5040.} \end{array}$$

&c.

Page 5 [Folio 112]

Notation: Lower case letters a, b, c, d, f, and g head each column. Perhaps Harriot did not use e because for him (as for Viète) vowels were reserved for unknown quantities, though he did use a for the constant difference, which has special status. The symbol Δ indicates an increasing column, also sometimes indicated by c for *crescente*. The symbol ∇ indicates a decreasing column, also sometimes indicated by d for *decrescente*. (This double use of c and d does not in fact cause any confusion). A small square $\square$ or the letter ε is used for columns of equal entries.

Pages 5 to 7 contain difference tables and formulae for their entries. The columns of these tables are the (general) arithmetic progressions of the title of the treatise.

At the top of page 5 are two tables of differences. In each case column g may be taken as the starting point. Column f contains successive differences between entries in column g, column d contains second differences, and so on. Eventually we reach a column of constant fifth differences, headed a.

The notation (together with the table that follows) suggests that Harriot actually devised his examples by *starting* with the constant difference in column a, along with the first entries for each of columns b to g, then filling in the entries by addition. Harriot's arrangement of the two tables so that their outlines are mirror images of one another may have been to demonstrate that such tables may be formed in either direction.

The main central table begins with a constant difference a and builds up successive entries algebraically, using only the first entries in each of the subsequent columns b to g. Note that a, b, c, d, f, and g are now the actual first entries of the columns rather than simply column headings. Harriot has drawn a diagonal line under the entries required to give at least one entry in the constant difference column.

In the lower table, Harriot wrote a general formula for the $(n + 1)$th entry in each column of the difference table, using the formulae for the triangular number coefficients already established on page 3. The third-column formula, for instance, reads

$$c + \frac{n}{1}b + \frac{(n-1)n}{1 \cdot 2}a.$$

The two small inset charts to the right of the middle table summarize the patterns in the columns of the difference tables. The first shows that all columns (except the first) are increasing; the second shows the pattern of signs in each column. The use of these charts will become more apparent on page 6, where successive entries of some columns will be obtained by subtraction rather than addition.

Related material: BL Add MSS 6782, ff. 211v–212, 217, 348; 6784, f. 207–207v; 6786, f. 417v.

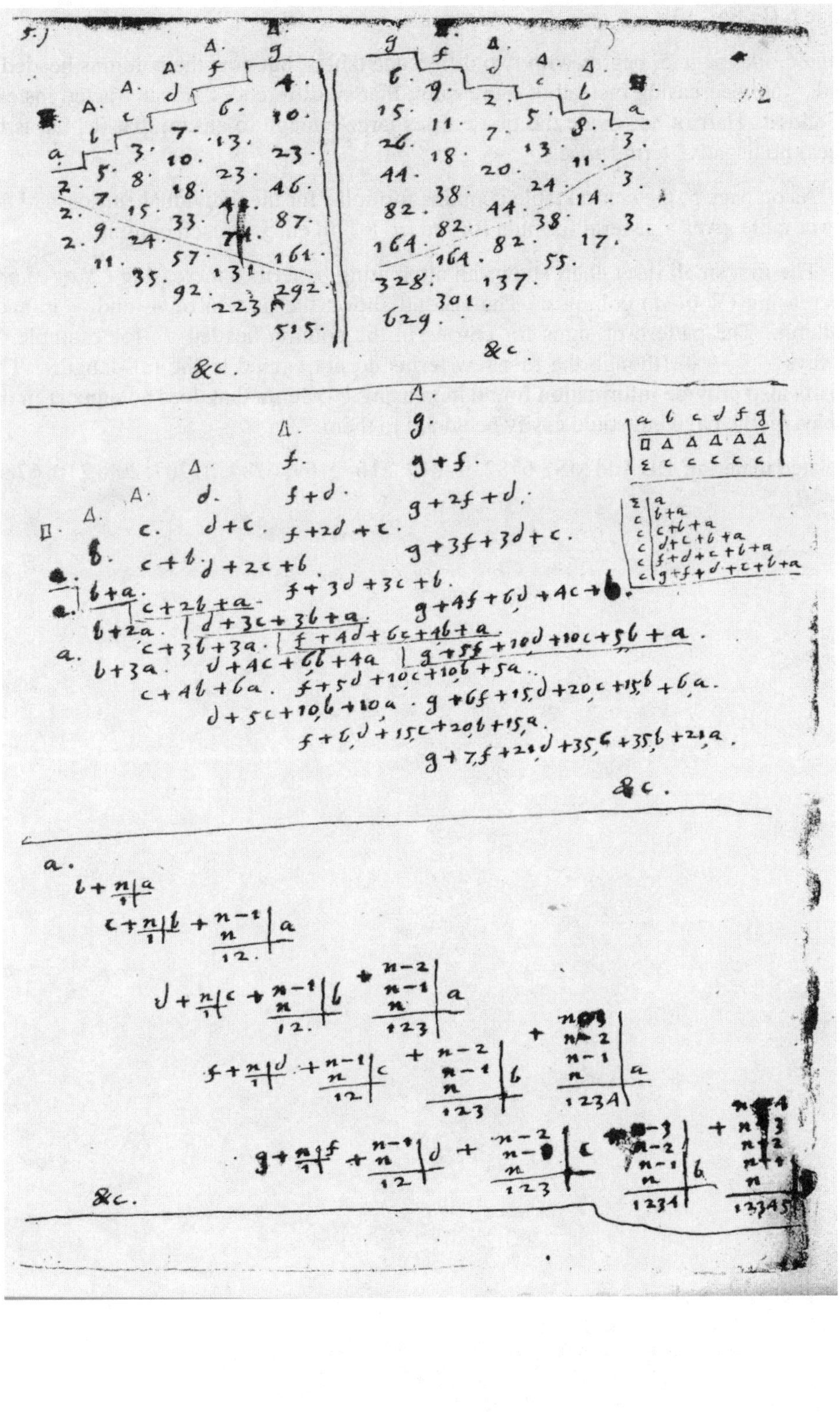

Page 6 [Folio 113]

Page 6, like page 5, begins with two difference tables, but now the columns headed c and f are decreasing instead of increasing; that is, differences are subtracted instead of added. Harriot has made the first entries large enough to ensure that (as far as he goes) no negative terms arise.

As on page 5, the central table contains formulae for the individual entries, and the lower table gives a general formula for the $(n+1)$th entry of each column.

The first small inset chart shows an alternating pattern of increasing (Δ or c) and decreasing (∇ or d) columns. The second shows the pattern of $+$ and $-$ in each column. The pattern of signs for entries in the column headed f, for example, is always $--++$ (though the first few terms do not extend to the full length). The charts also provide information for an increasing g column that does not appear in the tables on the page but could easily be added to them.

Related material: BL Add MSS 6782, ff. 213, 216–216v; 6784, ff. 207, 208, 210; 6786, f. 410.

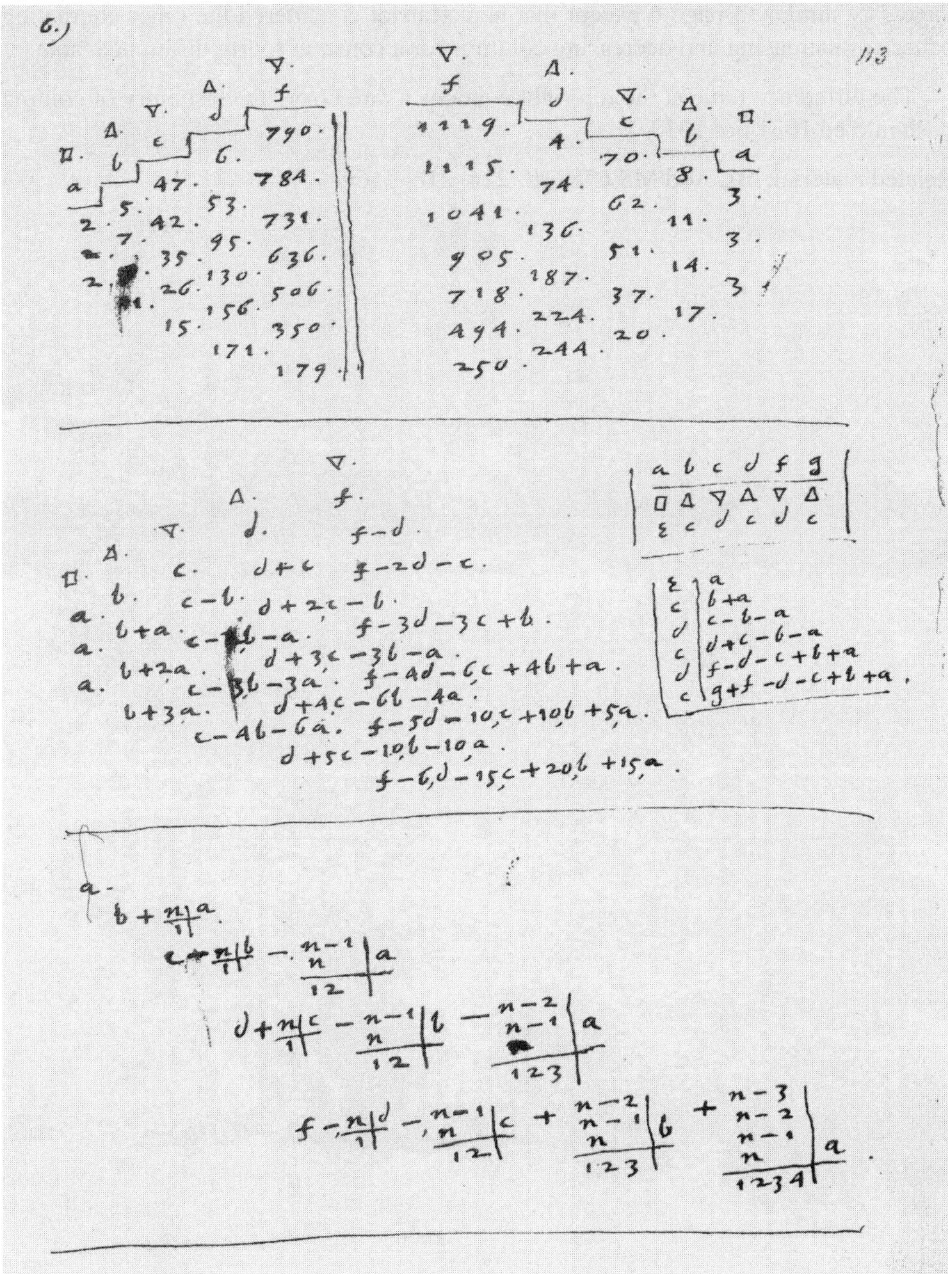

Page 7 [Folio 114]

Page 7 is similar to page 6 except that here Harriot considered the other alternating pattern of increasing and decreasing columns for a constant fourth difference table.

The difference table at the top right contains a rare error: the last entry of column f should be 1031 not 1030.

Related material: BL Add MS 6782, ff. 214, 216–216v.

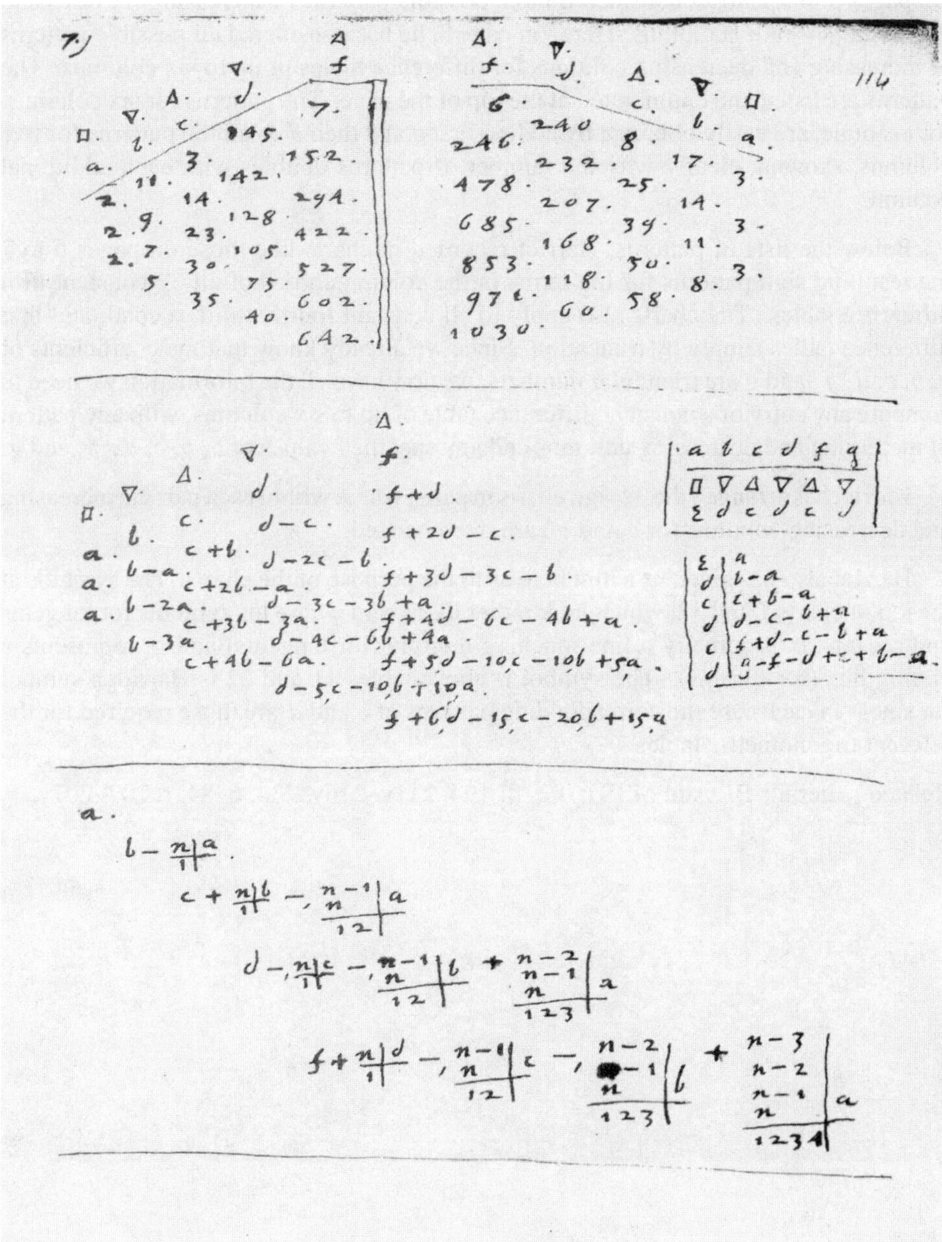

Page 8 [Folio 115]

On pages 5 to 7 Harriot considered three possible patterns of increasing (Δ or c) and decreasing (∇ or d) columns. Here, on page 8, he has considered all possible patterns of increasing and decreasing columns for difference tables of up to six columns. The patterns are listed and enumerated at the top of the page. The patterns for six columns, for example, are easily obtained by adding first c and then d to all the patterns for five columns, showing clearly why the number of patterns doubles with each additional column.

Below the lists of patterns, Harriot recorded in charts like those on pages 5 to 7 the resulting sign patterns for the terms in the column entries of all 32 constant fifth difference tables. The charts also apply to all constant fourth, third, second, and first difference tables simply by truncation. Since we already know that the coefficients of a, b, c, d, f, and g are triangular numbers, we now have all the information we need to compute any entry of a constant difference table of up to six columns, with any pattern of increasing and decreasing columns and any specified values of a, b, c, d, f, and g.

Harriot has arranged the 32 sign charts in pairs, where within each pair the increasing and decreasing columns (or c and d) are interchanged.

Harriot also has given us a hint here as to the purpose of the charts. The symbols at the top of charts 1 and 32, which look rather like σ and ψ, are his symbols for tangents and secants; σ is actually a line touching a circle (for tangents) and ψ represents a cutting line (for secants). The symbol υ above tables 11 and 22 is Harriot's symbol for sines. In each case the corresponding patterns of c and d are those required for the relevant trigonometric tables.

Related material: BL Add MSS 6782, ff. 194, 211v–216v, 234; 6784, f. 207–207v.

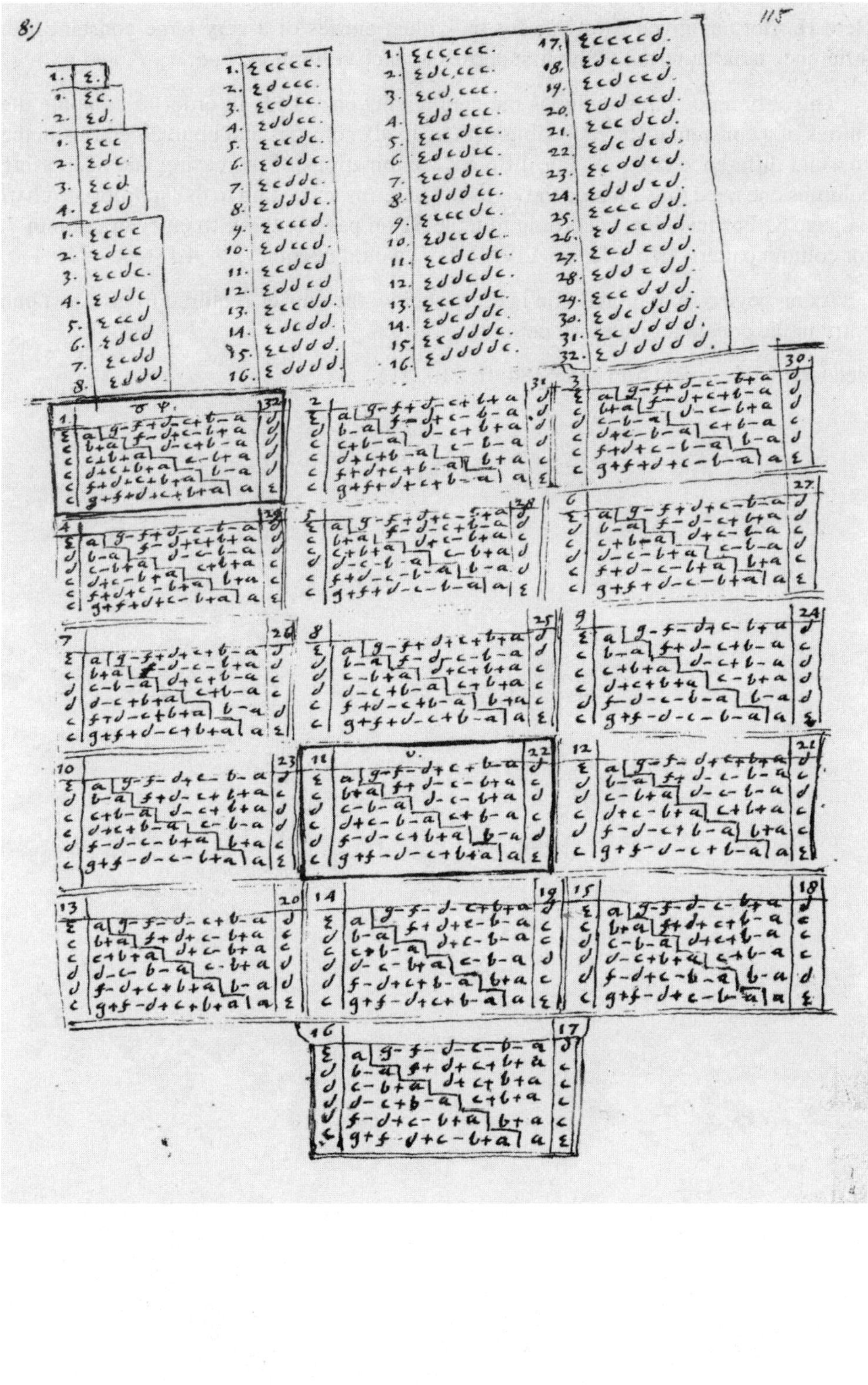

Page 9 [Folio 116]

Here Harriot has given formulae for individual entries of a very large constant fifth difference table in terms of the first entries of each column, a, b, c, d, f, and g.

This difference table includes the coefficients one needs in order to compute the entries of a constant difference table with up to six columns and up to 24 entries in the constant difference column. For different combinations of increasing and decreasing columns one need only change the signs of the terms according to the appropriate chart on page 8. For instance, according to table 32 on page 8, the fifth entry of column f for column pattern $\varepsilon ddddd$ (or □▽▽▽▽▽) would become $f - 4d + 6c - 4b + a$.

As on page 5, a diagonal line is drawn below the entries required to give just one entry in the constant difference column.

Related material: BL Add MS 6784, ff. 210, 211.

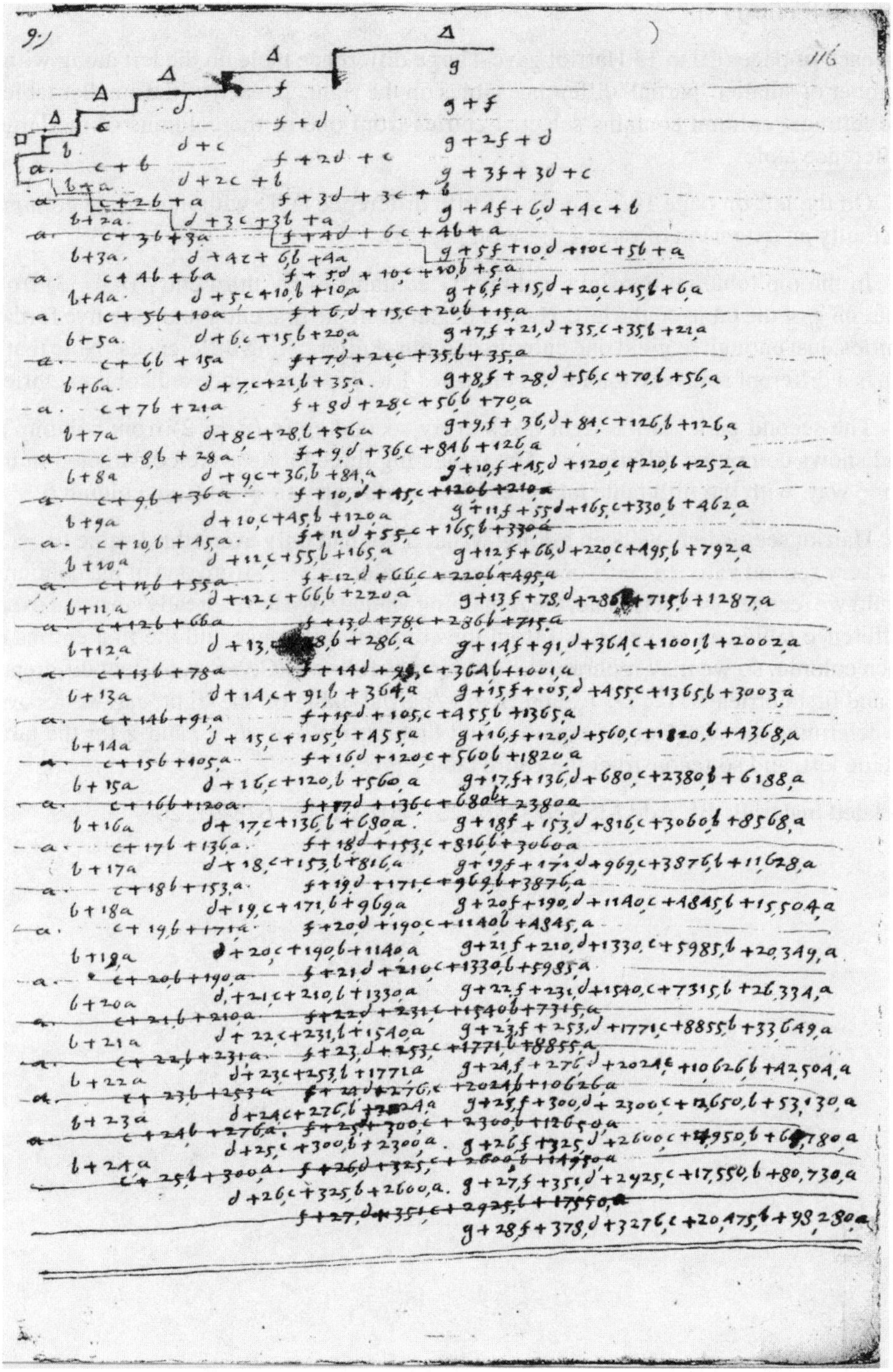

Page 10 [Folio 117]

On each of pages 10 to 13 Harriot gave a large difference table on the left along with a number of smaller 'partial' difference tables on the right. In each of the smaller tables, the leftmost column contains selected entries from one of the columns of the larger difference table.

On the left on page 10 is a constant fifth difference table with increasing columns (actually an extension of one of those on page 5).

In the top table on the right, column G contains every third entry ($n = 3$) from column g of the table on the left. Harriot began from the first entry and took five further entries, just enough to yield one entry in column A after taking differences. Note that n plays a different role here than it did on pages 1 to 7, where it indexed column entries.

The second table on the right takes every second entry ($n = 2$) from column f, and shows computed differences. The remaining three tables were constructed in the same way, with the fifth table taking every seventh entry ($n = 7$) from column b.

Harriot seems to have been asking: What if we had only every third value ($n = 3$) or every second value ($n = 2$) or every seventh value ($n = 7$) from one of the columns; could we recover, or interpolate, the remaining values? We have already seen that every difference table can be generated from the constant difference and the first entries of each column, so we may rephrase the question as follows: Given a constant difference A and first entries B, C, D, F, and G in a 'partial' table on the right, can we recover or determine the constant difference a and first entries b, c, d, f, and g for the table on the left, and so reconstruct the entire table?

Related material: BL Add MSS 6782, ff. 201–201v, 204v; 6784, f. 208.

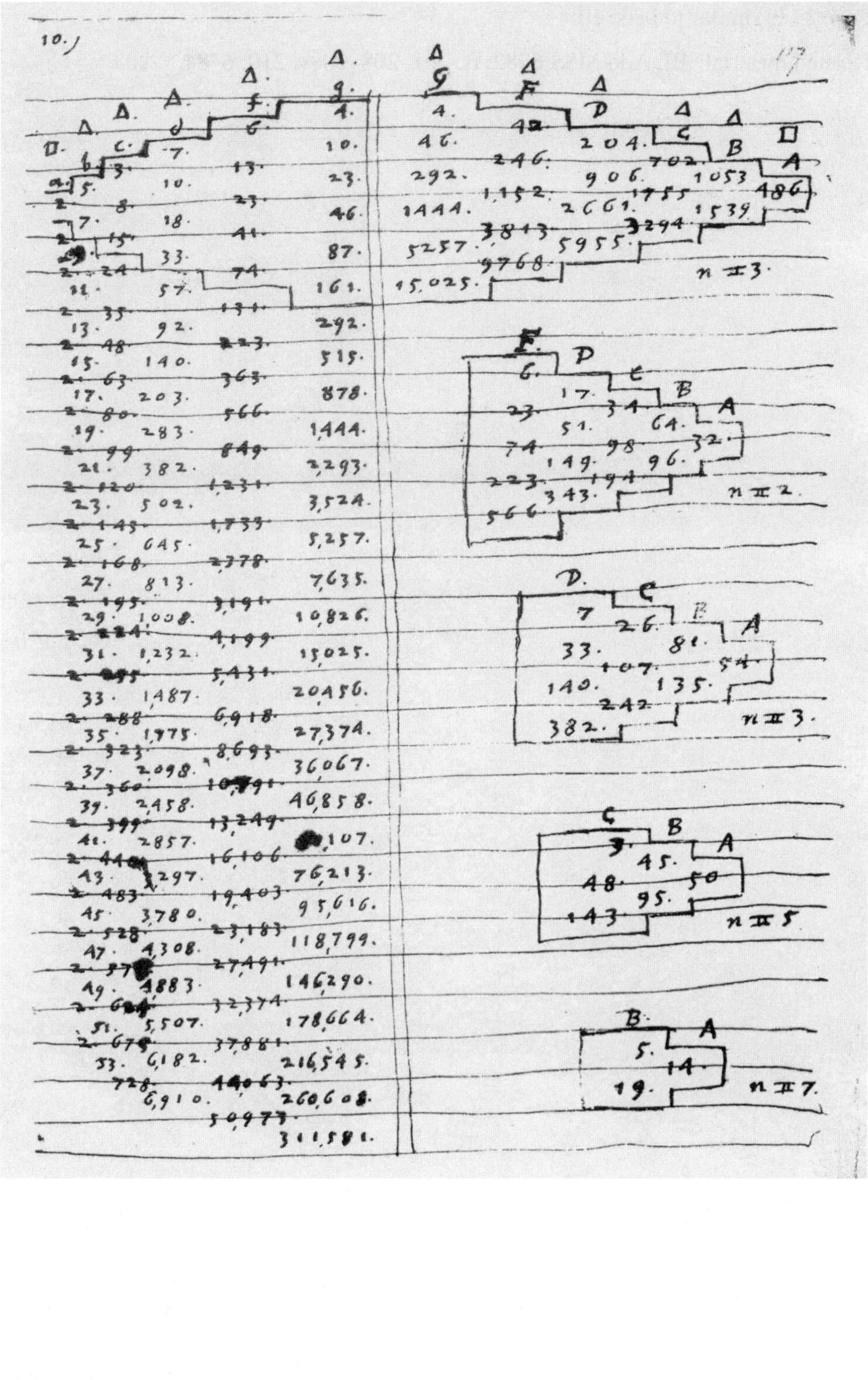

Page 11 [Folio 118]

Page 11 is similar to page 10.

Related material: BL Add MSS 6782, ff. 201, 208–208v, 210; 6784, f. 208.

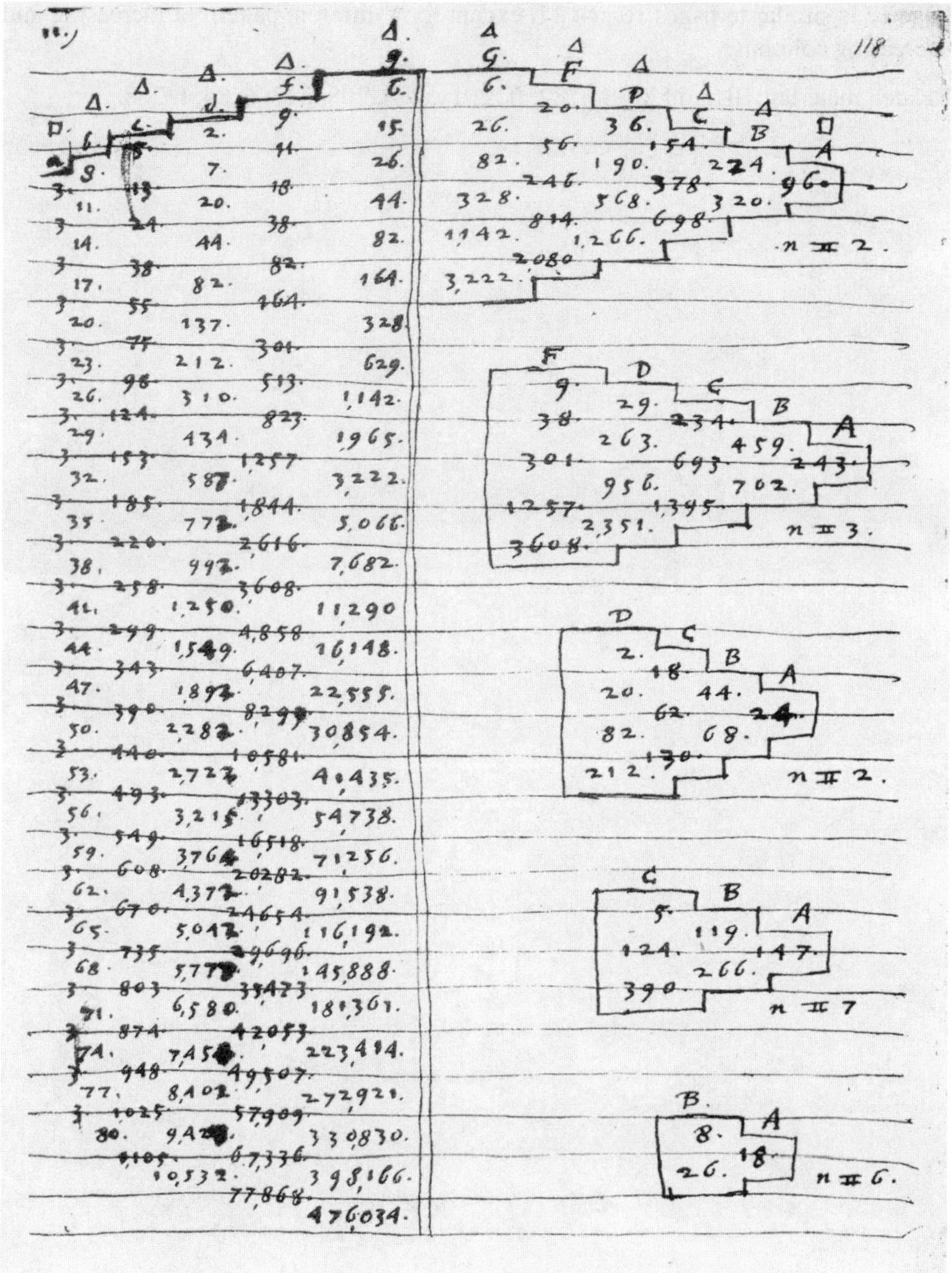

Page 12 [Folio 119]

Page 12 is similar to pages 10 and 11, except for a different pattern of increasing and decreasing columns.

Related material: BL Add MSS 6782, ff. 201v, 204, 208, 347; 6784, f. 208.

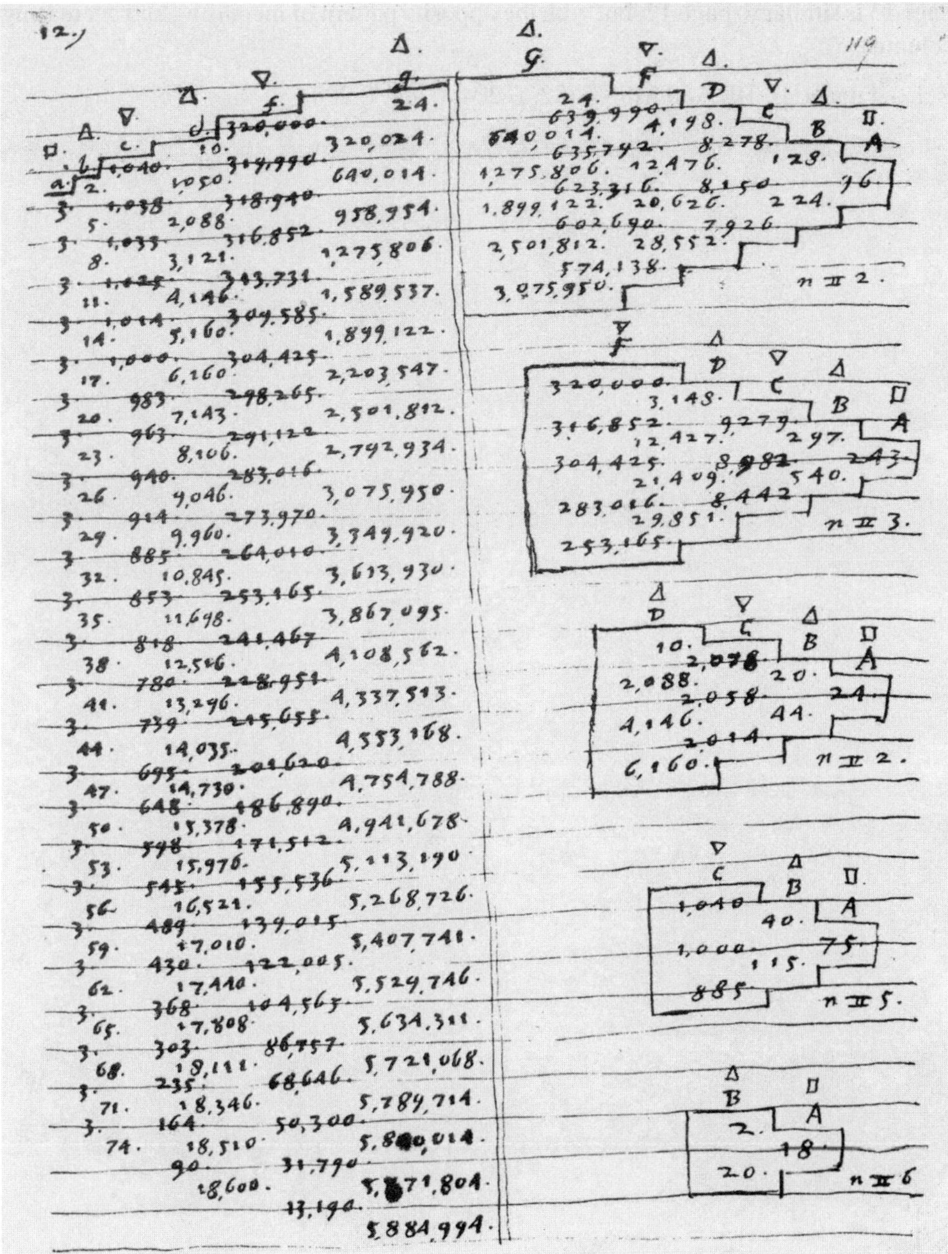

Page 13 [Folio 120]

Page 13 is similar to page 12, but with the opposite pattern of increasing and decreasing columns.

Related material: BL Add MSS 6782, f. 204; 6784, f. 208.

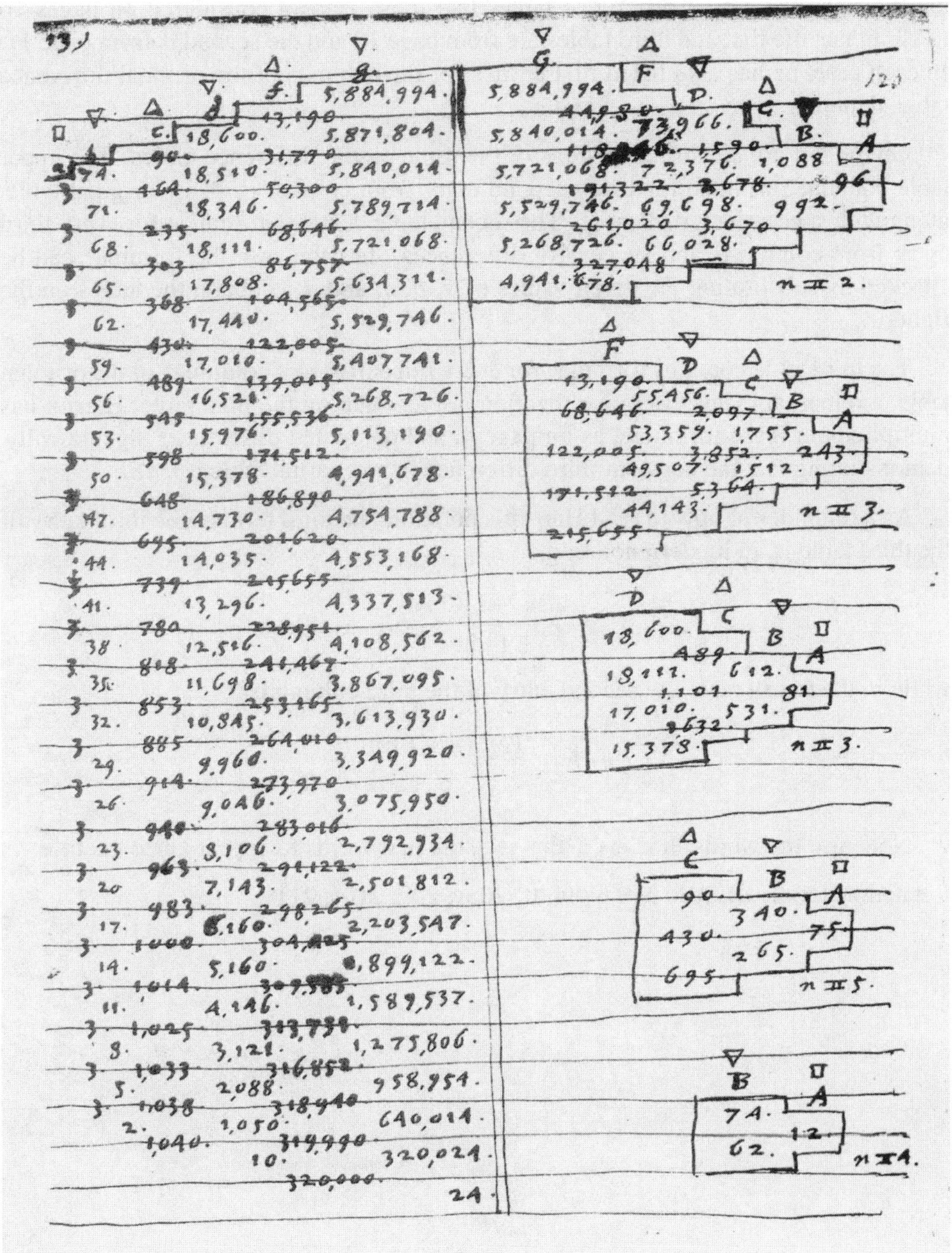

Page 14 [Folio 121]

On the right are partial difference tables like those Harriot considered on pages 10 to 13; in fact the first and third tables are from page 10 and the second is from page 11. In each case, he has also listed first entries for the columns of the original difference table, denoting them by a, b, c, and d.

On the left are algebraic versions of the same partial difference tables. The upper table contains formulae for every second entry from column d of the original table, along with successive differences. The second table contains formulae for every third entry from column d and successive differences. In each case the formulae can be checked by substituting the given values of a, b, c, and d to obtain the tables on the right.

The third table contains formulae for every nth entry from column d of the original table, a remarkable level of generalization. Below that, in the final table, Harriot has multiplied out these formulae, as on page 4, and computed differences algebraically, demonstrating that the constant third difference in any partial table is $nnna$.

As a reminder of how to read Harriot's notation, the third term of the third entry of the third table is, in modern notation,

$$\frac{(2n-1)(2n)}{1 \cdot 2} b,$$

while in the fourth table the second entry of the third column is

$$nnb + \frac{12nnn - 6nn}{6} a.$$

Note also the words 'hoc est' ('that is'), the first words to appear since the title.

Related material: BL Add MS 6784, ff. 208v, 209, 210v, 211v.

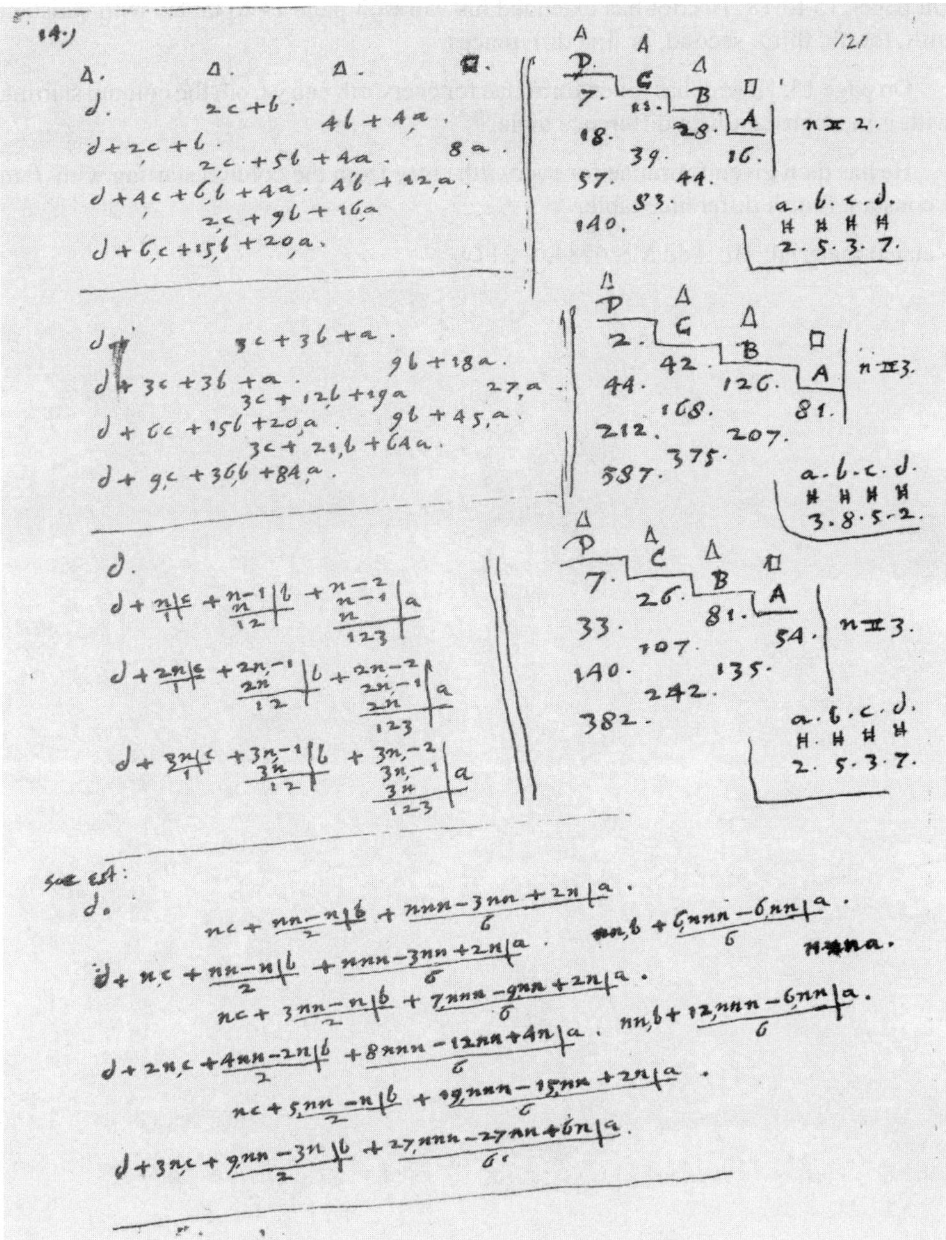
14.) /121

Δ. Δ. Δ. □.
d. 2c + b. 4b + 4a. 8a.
d + 2c + b. 2c + 5b + 4a. 4b + 12a.
d + 4c + 6b + 4a. 2c + 9b + 16a.
d + 6c + 15b + 20a.

D 7. 18. 57. 140. — C 11. 39. 83. — B 28. 44. — A 16. — n ∏ 2. — a. b. c. d. ∏ ∏ ∏ ∏ 2. 5. 3. 7.

d. 3c + 3b + a. 9b + 18a. 27a.
d + 3c + 3b + a. 3c + 12b + 19a. 9b + 45a.
d + 6c + 15b + 20a. 3c + 21b + 64a.
d + 9c + 36b + 84a.

D 2. 44. 212. 587. — C 42. 168. 375. — B 126. 207. — A 81. — n ∏ 3. — a. b. c. d. ∏ ∏ ∏ ∏ 3. 8. 5. 2.

d.
$d + \frac{n}{1}c + \frac{n-1 \cdot n}{1\,2}b + \frac{n-2 \cdot n-1 \cdot n}{1\,2\,3}a$
$d + \frac{2n}{1}c + \frac{2n-1 \cdot 2n}{1\,2}b + \frac{2n-2 \cdot 2n-1 \cdot 2n}{1\,2\,3}a$
$d + \frac{3n}{1}c + \frac{3n-1 \cdot 3n}{1\,2}b + \frac{3n-2 \cdot 3n-1 \cdot 3n}{1\,2\,3}a$

D 7. 33. 140. 382. — C 26. 107. 242. — B 81. 135. — A 54. — n ∏ 3. — a. b. c. d. ∏ ∏ ∏ ∏ 2. 5. 3. 7.

sive est:
d.
$nc + \frac{nn-n}{2}b + \frac{nnn-3nn+2n}{6}a$. $nnb + \frac{6nnn-6nn}{6}a$. $nnna$.
$d + nc + \frac{nn-n}{2}b + \frac{nnn-3nn+2n}{6}a$.
$nc + \frac{3nn-n}{2}b + \frac{7nnn-9nn+2n}{6}a$. $nnb + \frac{12nnn-6nn}{6}a$.
$d + 2nc + \frac{4nn-2n}{2}b + \frac{8nnn-12nn+4n}{6}a$.
$nc + \frac{5nn-n}{2}b + \frac{19nnn-15nn+2n}{6}a$.
$d + 3nc + \frac{9nn-3n}{2}b + \frac{27nnn-27nn+6n}{6}a$.

Page 15 [Folio 122]

On pages 15 to 18, Harriot has extended his work on page 14 to tables with constant fifth, fourth, third, second, or first differences.

On page 15, Harriot has given formulae for every nth entry from the column starting with g in a constant fifth difference table.

He has then given formulae for every nth entry from the column starting with f in a constant fourth difference table.

Related material: BL Add MS 6784, f. 211v.

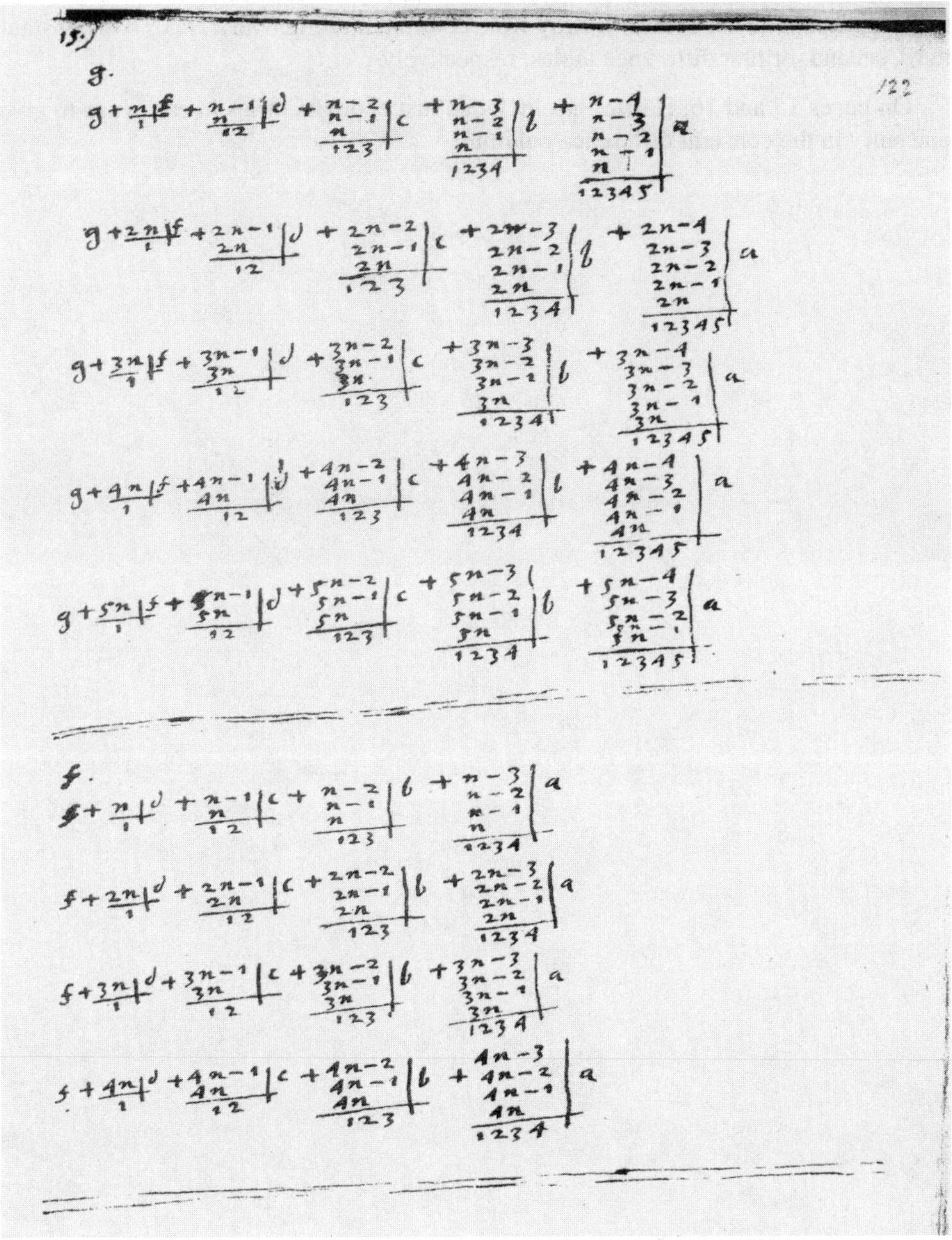

15.) 122

g.

$$g+\frac{n}{1}f+\frac{(n-1)n}{12}d+\frac{(n-2)(n-1)n}{123}c+\frac{(n-3)(n-2)(n-1)n}{1234}b+\frac{(n-4)(n-3)(n-2)(n-1)n}{12345}a$$

$$g+\frac{2n}{1}f+\frac{(2n-1)2n}{12}d+\frac{(2n-2)(2n-1)2n}{123}c+\frac{(2n-3)(2n-2)(2n-1)2n}{1234}b+\frac{(2n-4)(2n-3)(2n-2)(2n-1)2n}{12345}a$$

$$g+\frac{3n}{1}f+\frac{(3n-1)3n}{12}d+\frac{(3n-2)(3n-1)3n}{123}c+\frac{(3n-3)(3n-2)(3n-1)3n}{1234}b+\frac{(3n-4)(3n-3)(3n-2)(3n-1)3n}{12345}a$$

$$g+\frac{4n}{1}f+\frac{(4n-1)4n}{12}d+\frac{(4n-2)(4n-1)4n}{123}c+\frac{(4n-3)(4n-2)(4n-1)4n}{1234}b+\frac{(4n-4)(4n-3)(4n-2)(4n-1)4n}{12345}a$$

$$g+\frac{5n}{1}f+\frac{(5n-1)5n}{12}d+\frac{(5n-2)(5n-1)5n}{123}c+\frac{(5n-3)(5n-2)(5n-1)5n}{1234}b+\frac{(5n-4)(5n-3)(5n-2)(5n-1)5n}{12345}a$$

f.

$$f+\frac{n}{1}d+\frac{(n-1)n}{12}c+\frac{(n-2)(n-1)n}{123}b+\frac{(n-3)(n-2)(n-1)n}{1234}a$$

$$f+\frac{2n}{1}d+\frac{(2n-1)2n}{12}c+\frac{(2n-2)(2n-1)2n}{123}b+\frac{(2n-3)(2n-2)(2n-1)2n}{1234}a$$

$$f+\frac{3n}{1}d+\frac{(3n-1)3n}{12}c+\frac{(3n-2)(3n-1)3n}{123}b+\frac{(3n-3)(3n-2)(3n-1)3n}{1234}a$$

$$f+\frac{4n}{1}d+\frac{(4n-1)4n}{12}c+\frac{(4n-2)(4n-1)4n}{123}b+\frac{(4n-3)(4n-2)(4n-1)4n}{1234}a$$

Page 16 [Folio 123]

Page 16 is a continuation of page 15, and here Harriot has written formulae analogous to those on page 15 for every nth entry from columns starting with d, c, or b in constant third, second, or first difference tables, respectively.

On pages 15 and 16, Harriot has included just enough entries in each case to give one entry in the constant difference column.

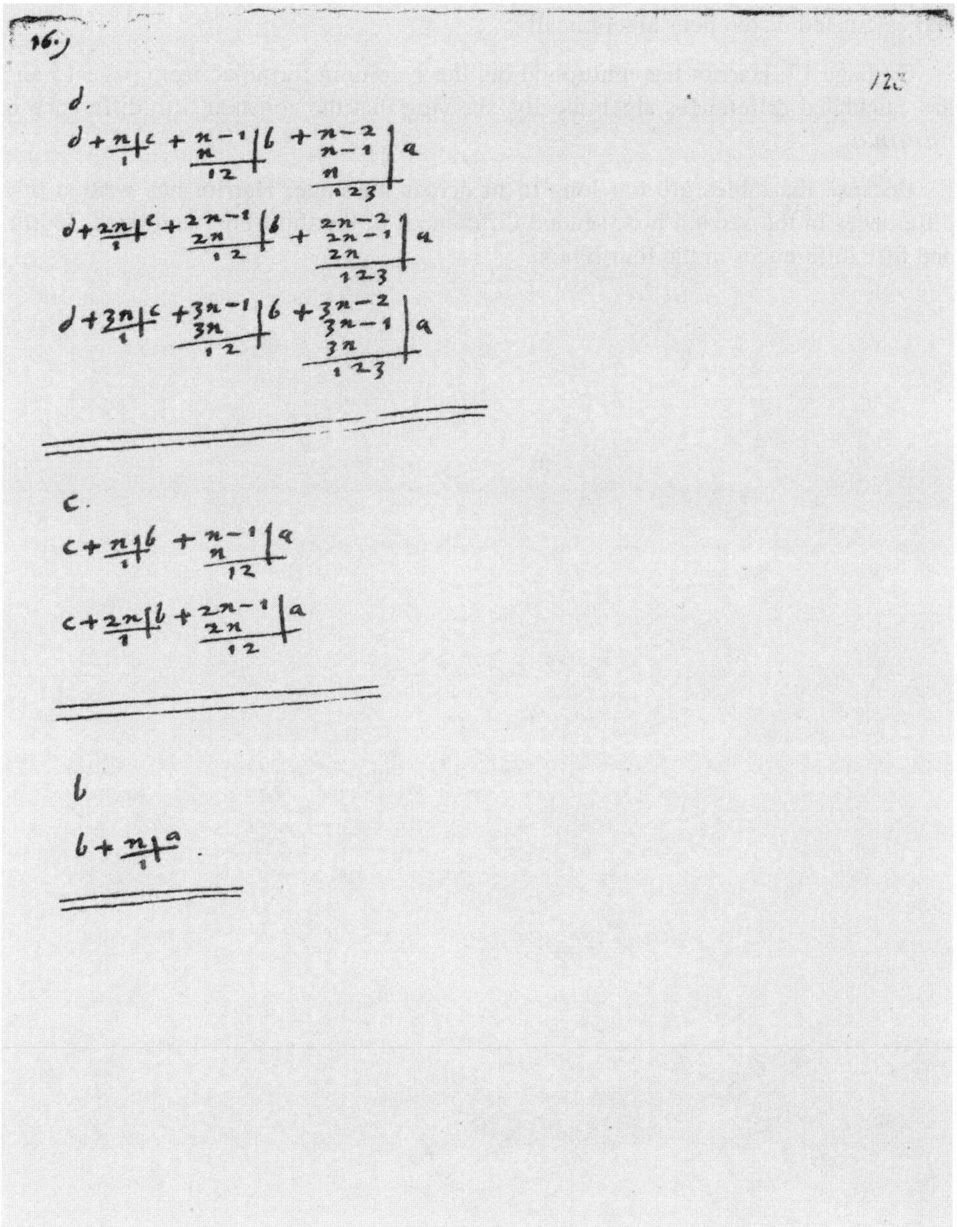

16.) 123

d.

$$d + \frac{n}{1}\Big|\, c + \frac{\substack{n-1\\ n}}{1\;2}\Big|\, b + \frac{\substack{n-2\\ n-1\\ n}}{1\;2\;3}\Big|\, a$$

$$d + \frac{2n}{1}\Big|\, c + \frac{\substack{2n-1\\ 2n}}{1\;2}\Big|\, b + \frac{\substack{2n-2\\ 2n-1\\ 2n}}{1\;2\;3}\Big|\, a$$

$$d + \frac{3n}{1}\Big|\, c + \frac{\substack{3n-1\\ 3n}}{1\;2}\Big|\, b + \frac{\substack{3n-2\\ 3n-1\\ 3n}}{1\;2\;3}\Big|\, a$$

c.

$$c + \frac{n}{1}\Big|\, b + \frac{\substack{n-1\\ n}}{1\;2}\Big|\, a$$

$$c + \frac{2n}{1}\Big|\, b + \frac{\substack{2n-1\\ 2n}}{1\;2}\Big|\, a$$

b

$$b + \frac{n}{1}\Big|\, a.$$

Page 17 [Folio 124]

On pages 17 and 18, Harriot has multiplied out the formulae from pages 15 and 16, and calculated differences algebraically.

On page 17, Harriot has multiplied out the g column formulae from page 15 and has calculated differences algebraically, showing that the constant fifth difference is $nnnnn,a$.

Because the tables are too long to fit across the page, Harriot has written first differences in the second box, second differences in the third box, and third, fourth, and fifth differences in the fourth box.

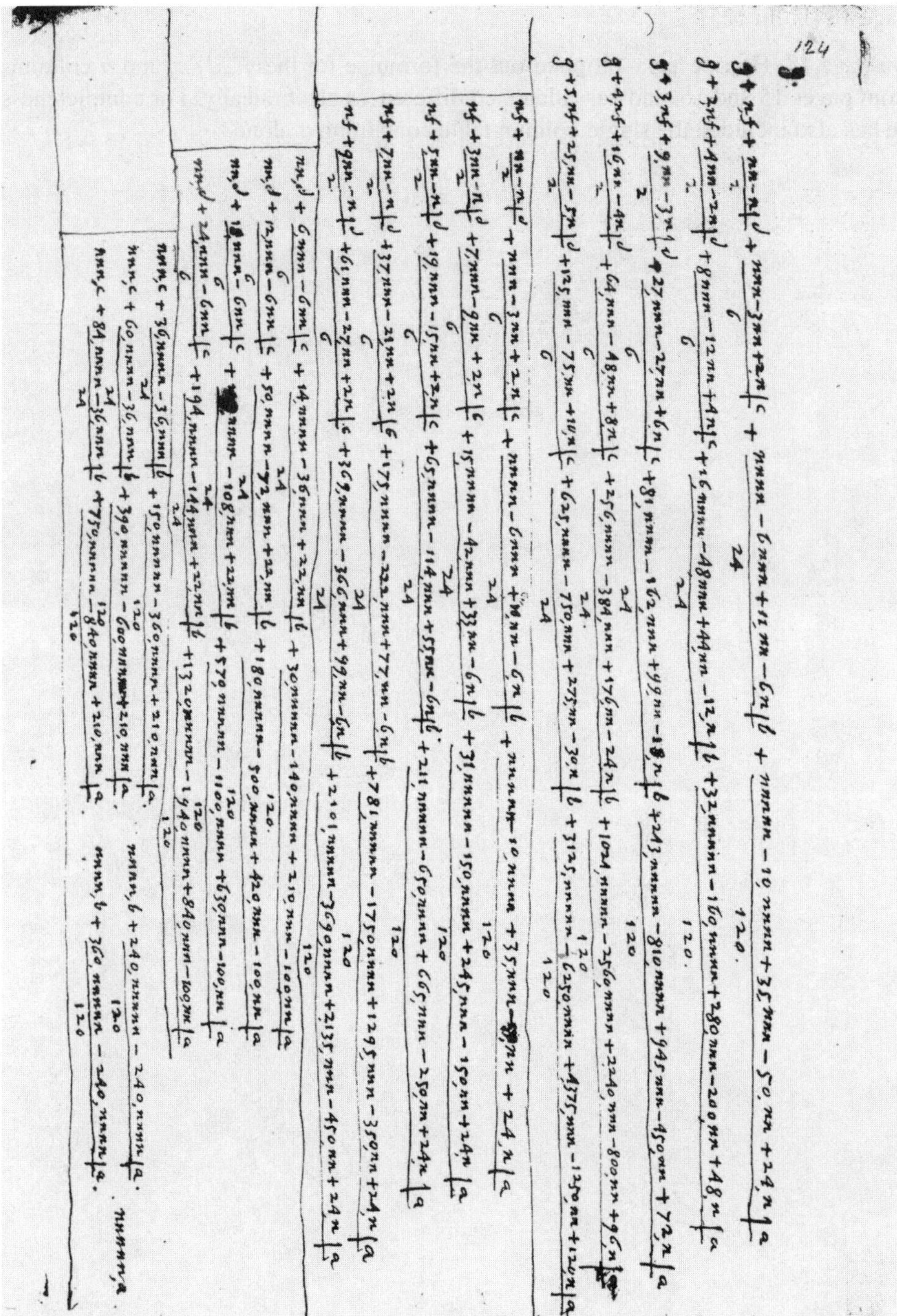

Page 18 [Folio 125]

On page 18, Harriot has multiplied out the formulae for the f, d, c, and b columns from pages 15 and 16, and has calculated differences algebraically. For completeness he has also included the single-column table containing a alone.

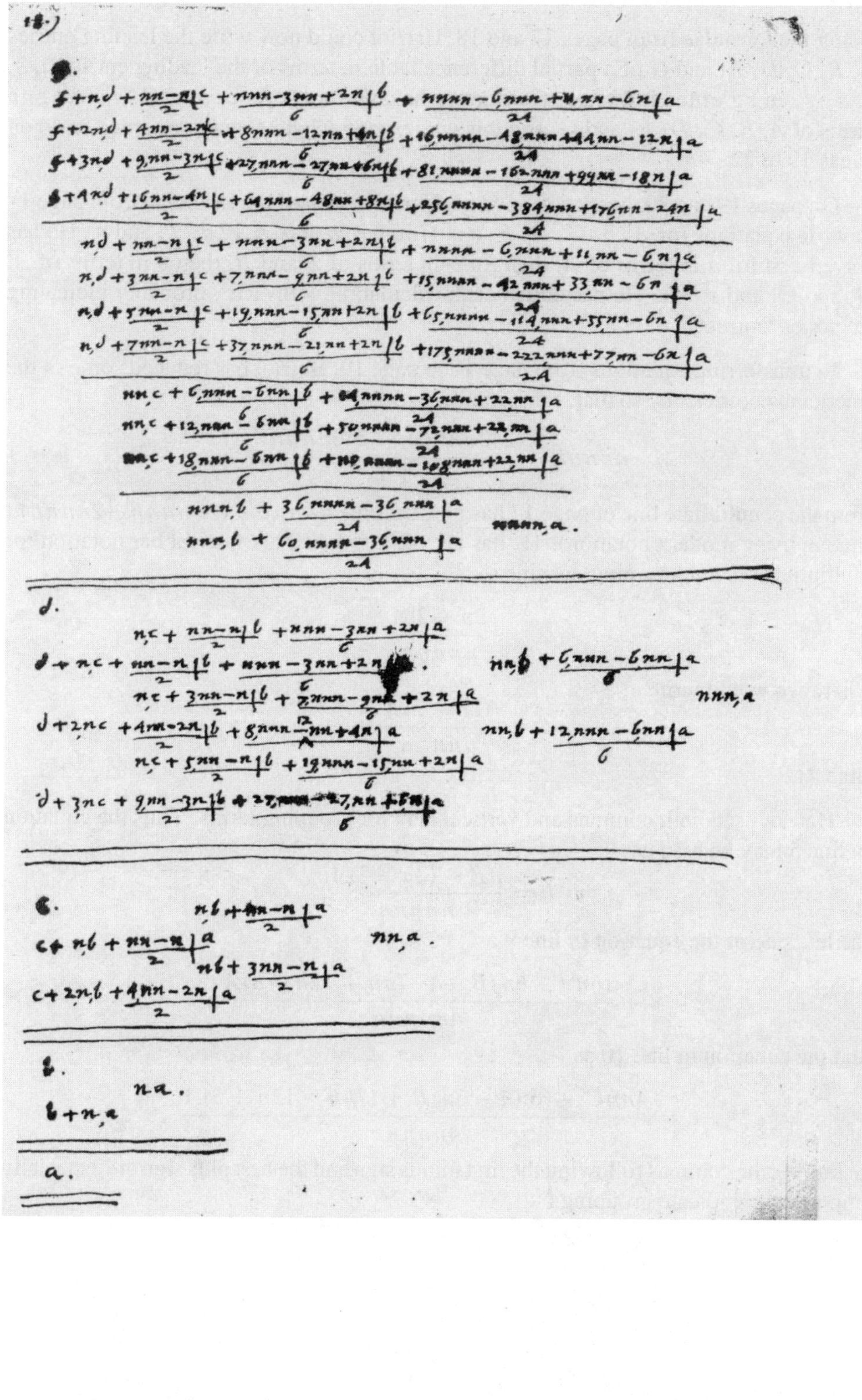

Page 19 [Folio 126]

Using the formulae from pages 17 and 18, Harriot could now write the leading entries A, B, C, D, F, and G of a partial difference table in terms of the leading entries a, b, c, d, f, and g of the full table. Conversely he could solve for a, b, c, d, f, and g in terms of A, B, C, D, F, and G, and this was exactly what he now proceeded to do on pages 19 to 22.

On pages 19 and 20, headed 'Canon g', Harriot has used the formulae from page 17 to write equations for A, B, C, D, F, and G in terms of a, b, c, d, f, and g. He has solved first for a in terms of A, then for b in terms of A and B, then c in terms of A, B, and C, and so on. He has ended each calculation with 'RE', probably indicating 'recto' or 'correct'.

In transferring equations from page 17 to page 19, Harriot has reduced some of the fractional coefficients, so that, for instance,

$$nnnnb + \frac{240nnnnnn - 240nnnnn}{120}a$$

from the penultimate line of page 17 has now become $nnnnb + (2nnnnn - 2nnnn)a$ (line 3, using modern notation). He has also reintroduced his vertical bar notation for multiplication and grouping, writing

$$\frac{2n - 2|A}{nnnnn}$$

where we would write

$$\frac{(2n - 2)A}{nnnnn}$$

(line 4).

Harriot used both commas and vertical bars for grouping terms. Thus the equation in line 5 may be read as

$$b = \frac{nB - (2n - 2)A}{nnnnn},$$

the left side of the equation in line 9 as

$$c + \frac{(6nn - 6n)B + (-7nn + 12n - 5)A}{4nnnnn},$$

and the equation in line 10 as

$$c = \frac{4nnC - (6nn - 6n)B + (7nn - 12n + 5)A}{4nnnnn}.$$

In line 10, the commas following the first minus sign and the first plus sign are especially important to indicate grouping.

Related material: BL Add MS 6782, ff. 211–211v.

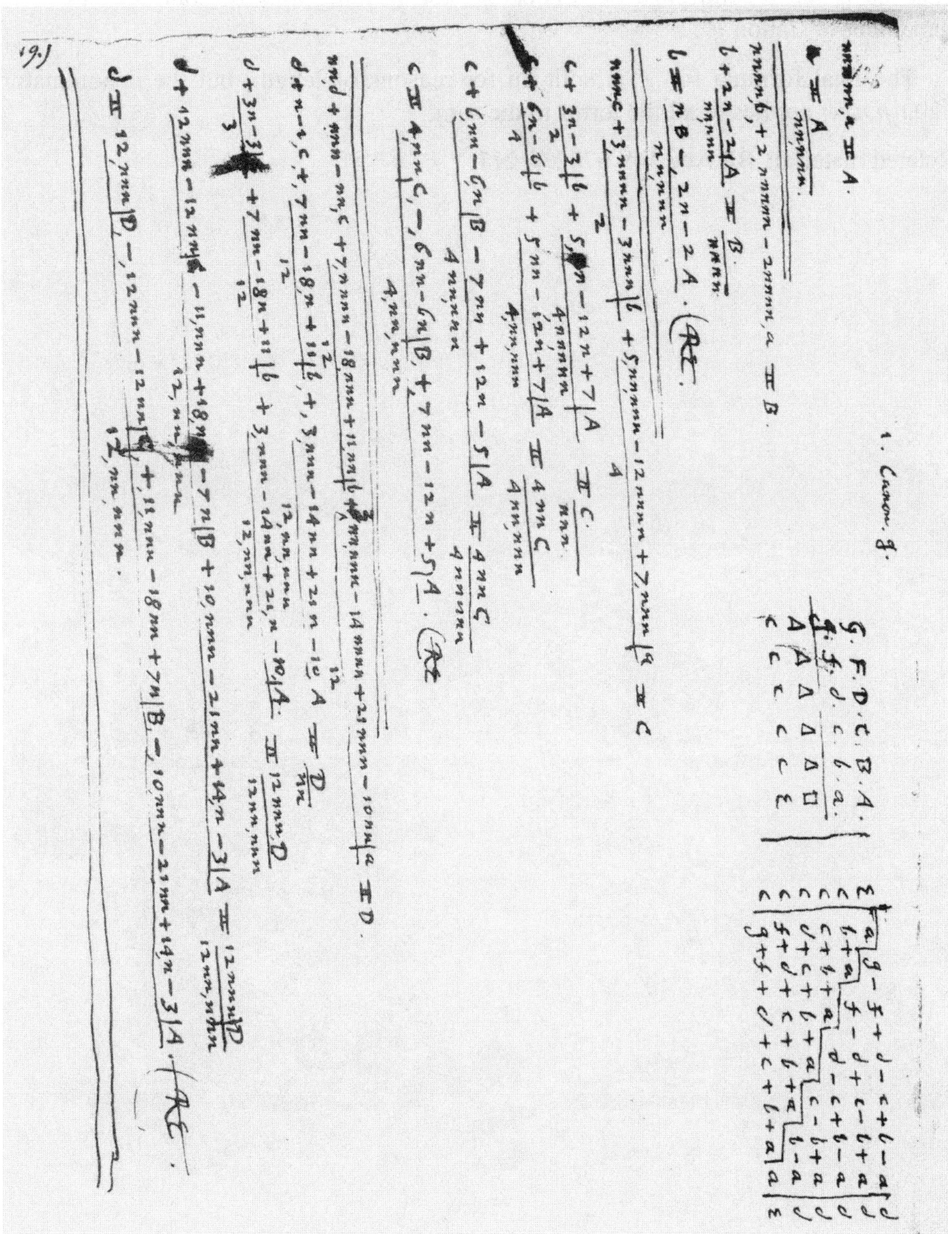

Page 20 [Folio 127]

Page 20 is the second page of 'Canon g'; the heading may be translated as 'The remainder of Canon g'.

The final formula for f is split up for reasons of length, but the denominator $120,nn,nnn$ applies to all the terms in the sum.

Related material: BL Add MS 6782, f. 211.

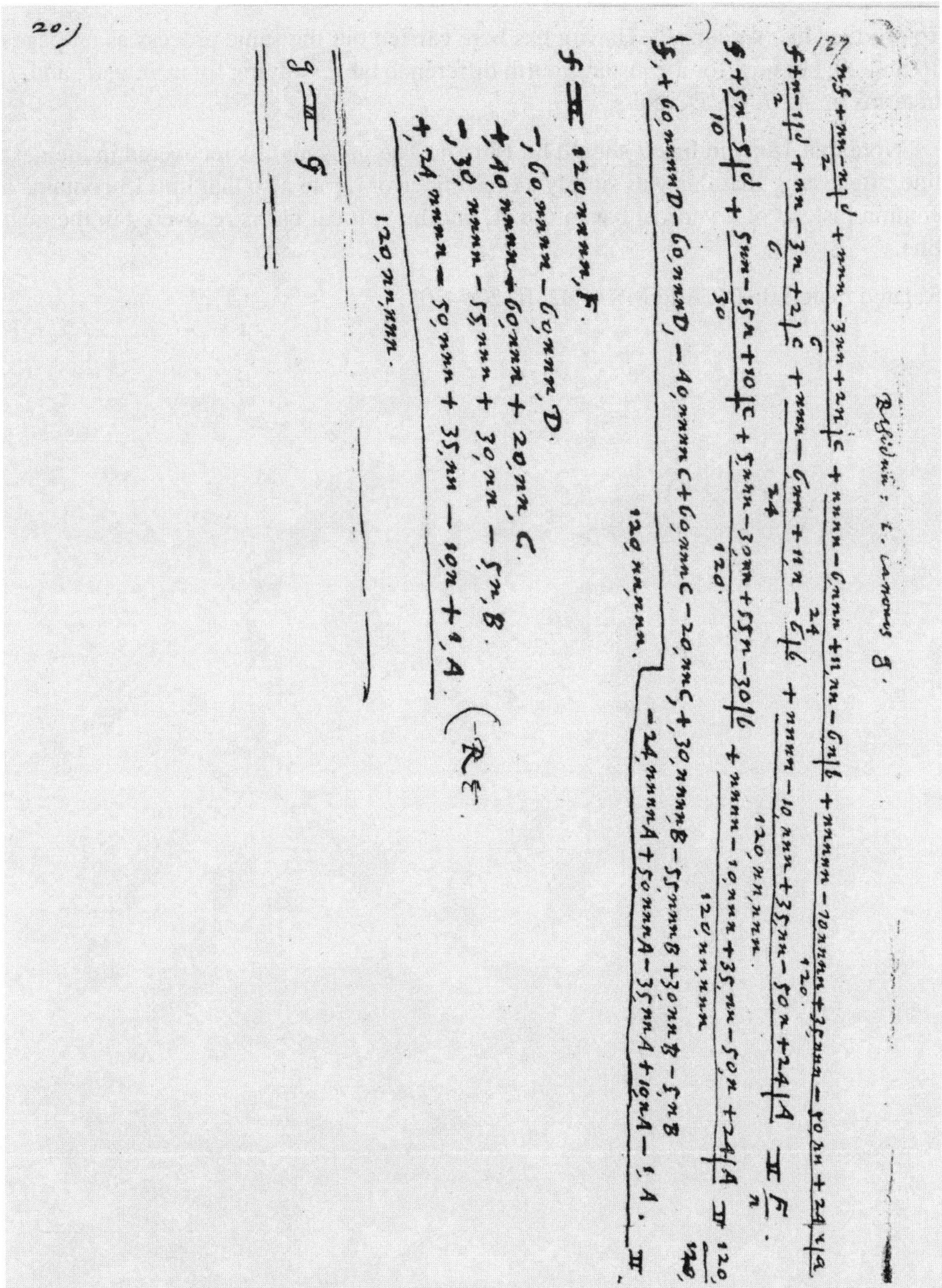

Page 21 [Folio 128]

To produce his 'Canon f', Harriot has here carried out the same process as on pages 19 and 20, but now for a constant fourth difference table, solving for a, b, c, d, and f in terms of A, B, C, D, and F.

Note that $18nn$ in line 7 should be $18nnn$. The missing n is recovered in the next line, suggesting that this was simply a copying error. Note also that line 13 contains a comma instead of a vertical bar in $6,n,A$, but the vertical bar is recovered in the next line.

Related material: BL Add MS 6782, ff. 206–207.

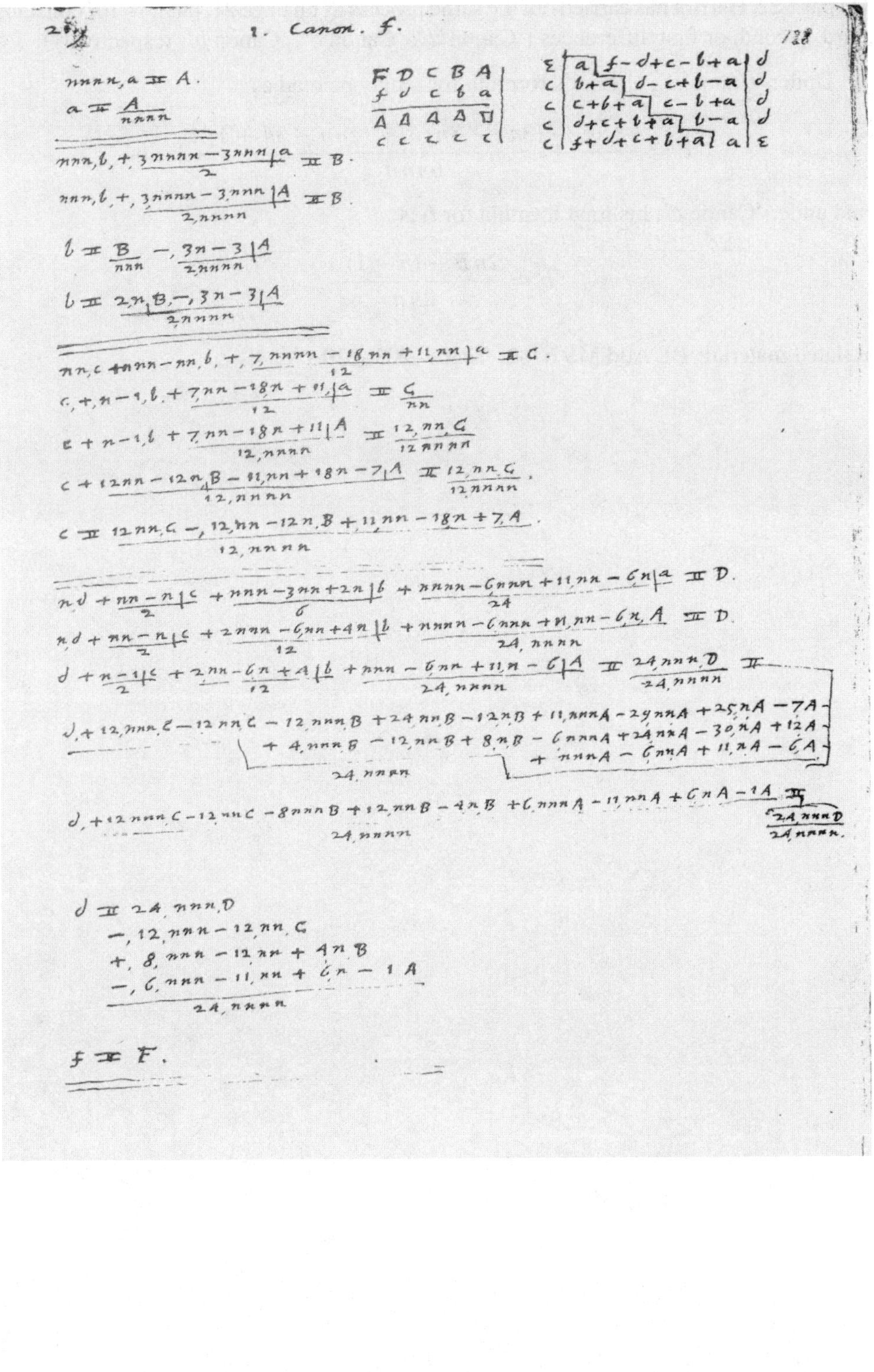

1. Canon. f.

128

$nnnn\,a = A.$

$a = \frac{A}{nnnn}$

F	D	C	B	A
f	d	c	b	a
Δ	Δ	Δ	Δ	◻
c	c	c	c	c

ε	a	f − d + c − b + a	d
c	b + a	d − c + b − a	d
c	c + b + a	c − b + a	d
c	d + c + b + a	b − a	d
c	f + d + c + b + a	a	ε

$nnn\,b + \frac{3nnnn - 3nnn}{2}\,a = B.$

$nnn\,b + \frac{3nnnn - 3nnn}{2\,nnnn}\,A = B.$

$b = \frac{B}{nnn} - \frac{3n - 3}{2\,nnnnn}\,A$

$b = \frac{2n\,B - (3n - 3)\,A}{2\,nnnn}$

$nn\,c + nnn - nn\,b + \frac{7\,nnnn - 18\,nn + 11\,nn}{12}\,a = C.$

$c + n - 1\,b + \frac{7\,nn - 18\,n + 11}{12}\,a = \frac{C}{nn}$

$c + n - 1\,b + \frac{7\,nn - 18\,n + 11}{12\,nnnn}\,A = \frac{12\,nn\,C}{12\,nnnn}$

$c + \frac{12\,nn - 12\,n\,B - 11\,nn + 18\,n - 7\,A}{12\,nnnn} = \frac{12\,nn\,C}{12\,nnnn}.$

$c = \frac{12\,nn\,C - (12\,nn - 12\,n)\,B + (11\,nn - 18\,n + 7)\,A}{12\,nnnn}.$

$n\,d + \frac{nn - n}{2}\,c + \frac{nnn - 3\,nn + 2\,n}{6}\,b + \frac{nnnn - 6\,nnn + 11\,nn - 6\,n}{24}\,a = D.$

$n\,d + \frac{nn - n}{2}\,c + \frac{2\,nnn - 6\,nn + 4\,n}{12}\,b + \frac{nnnn - 6\,nnn + 11\,nn - 6\,n\,A}{24\,nnnn} = D.$

$d + \frac{n - 1}{2}\,c + \frac{2\,nn - 6\,n + 4}{12}\,b + \frac{nnn - 6\,nn + 11\,n - 6}{24\,nnnn}\,A = \frac{24\,nnn\,D}{24\,nnnn} =$

$d + \frac{12\,nnn\,C - 12\,nn\,C - 12\,nnn\,B + 24\,nn\,B - 12\,n\,B + 11\,nnn\,A - 29\,nn\,A + 25\,n\,A - 7\,A + 4\,nnn\,B - 12\,nn\,B + 8\,n\,B - 6\,nnn\,A + 24\,nn\,A - 30\,n\,A + 12\,A + nnn\,A - 6\,nn\,A + 11\,n\,A - 6\,A}{24\,nnnn}$

$d + \frac{12\,nnn\,C - 12\,nn\,C - 8\,nnn\,B + 12\,nn\,B - 4\,n\,B + 6\,nnn\,A - 11\,nn\,A + 6\,n\,A - 1\,A}{24\,nnnn} = \frac{24\,nnn\,D}{24\,nnnn}.$

$d = \frac{24\,nnn\,D - (12\,nnn - 12\,nn)\,C + (8\,nnn - 12\,nn + 4\,n)\,B - (6\,nnn - 11\,nn + 6\,n - 1)\,A}{24\,nnnn}$

$f = F.$

Page 22 [Folio 129]

On page 22, Harriot has carried out the same process as on page 21, but now for constant third, second, or first differences (‘Canon d’, ‘Canon c’, ‘Canon b’, respectively).

Under ‘Canon d’, Harriot’s formula for c may be read as

$$c = \frac{6nnC - (3nn - 3n)B + (2nn - 3n + 1)A}{6nnn},$$

and under ‘Canon c’, his final formula for b is

$$b = \frac{2nB - (n - 1)A}{2nn}.$$

Related material: BL Add MS 6782, ff. 201–203, 220.

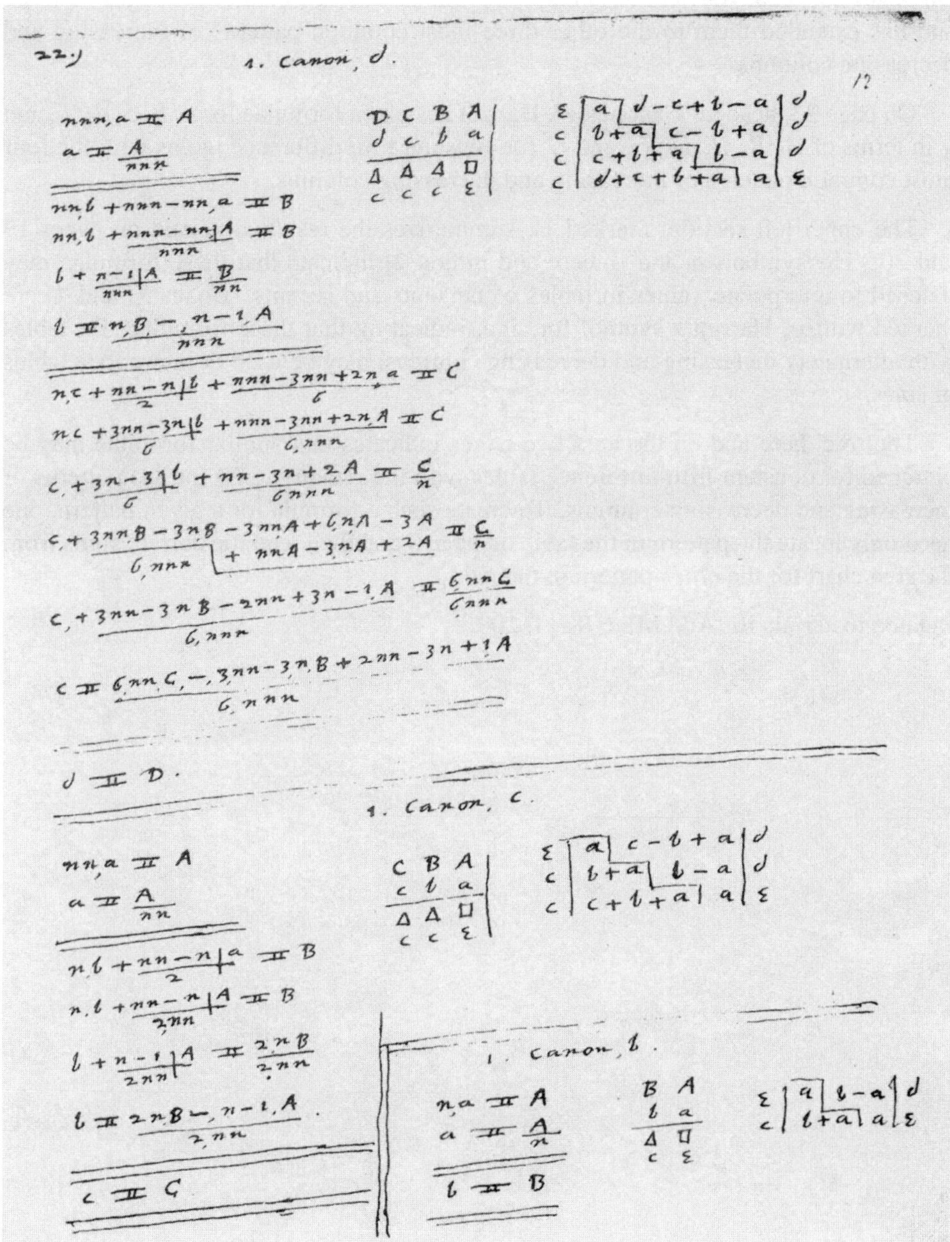

22.) 1. Canon, d. 12

$nnn,a = A$

$a = \frac{A}{nnn}$

$nn,b + nnn - nn,a = B$

$nn,b + \frac{nnn-nn}{nnn}A = B$

$b + \frac{n-1}{nnn}A = \frac{B}{nn}$

$b = \frac{n,B -, n-1,A}{nnn}$

$n,c + \frac{nn-n}{2}b + \frac{nnn-3nn+2n}{6}a = C$

$nc + \frac{3nn-3n}{6}b + \frac{nnn-3nn+2n,A}{6,nnn} = C$

$c, + \frac{3n-3}{6}b + \frac{nn-3n+2A}{6nnn} = \frac{C}{n}$

$c, + \frac{3nnB - 3nB - 3nnA + 6nA - 3A + nnA - 3nA + 2A}{6,nnn} = \frac{C}{n}$

$c, + \frac{3nn-3n\,B - 2nn+3n-1,A}{6,nnn} = \frac{6nnC}{6nnn}$

$c = \frac{6,nn,C, -, 3nn-3n,B + 2nn-3n+1A}{6,nnn}$

$d = D.$

D	C	B	A
d	c	b	a
Δ	Δ	Δ	□
c	c	c	ε

ε	a	d − c + b − a	d
c	b + a	c − b + a	d
c	c + b + a	b − a	d
c	d + c + b + a	a	ε

1. Canon, c.

$nn,a = A$

$a = \frac{A}{nn}$

$n,b + \frac{nn-n}{2}a = B$

$n,b + \frac{nn-n}{2nn}A = B$

$b + \frac{n-1}{2nn}A = \frac{2nB}{2nn}$

$b = \frac{2nB -, n-1,A}{2,nn}.$

$c = C$

C	B	A
c	b	a
Δ	Δ	□
c	c	ε

ε	a	c − b + a	d
c	b + a	b − a	d
c	c + b + a	a	ε

1. Canon, b.

$n,a = A$

$a = \frac{A}{n}$

$b = B$

B	A
b	a
Δ	□
c	ε

ε	a	b − a	d
c	b + a	a	ε

Page 23 [Folio 130]

On pages 23 to 25, Harriot has summarized the results he obtained on pages 19 to 22 and has extended them to the other three most common patterns of increasing and decreasing columns.

On page 23, headed 'Canones g', Harriot has given formulae for a, b, c, d, f, and g in terms of A, B, C, D, F, and G for constant fifth difference tables with the four most common patterns of increasing and decreasing columns.

The upper left section, marked 1), summarizes the results obtained on pages 19 and 20. The symbols σ and ψ here and in box 2) indicate that these formulae may be used to interpolate values in tables of tangents and secants. Boxes 3) and 4) are marked with υ, Harriot's symbol for sine, indicating that these formulae, for tables with alternately increasing and decreasing columns, may be used to interpolate tables of sines.

The '&c' here and on the next two pages indicates that similar formulae may be obtained for constant fifth difference tables with the remaining 28 (of 32) patterns of increasing and decreasing columns. To create such a formula for a given pattern, one need only locate the pattern in the table on page 8 and then read the correct signs from the sign chart for the *other* pattern in the pair.

Related material: BL Add MS 6782, f. 200.

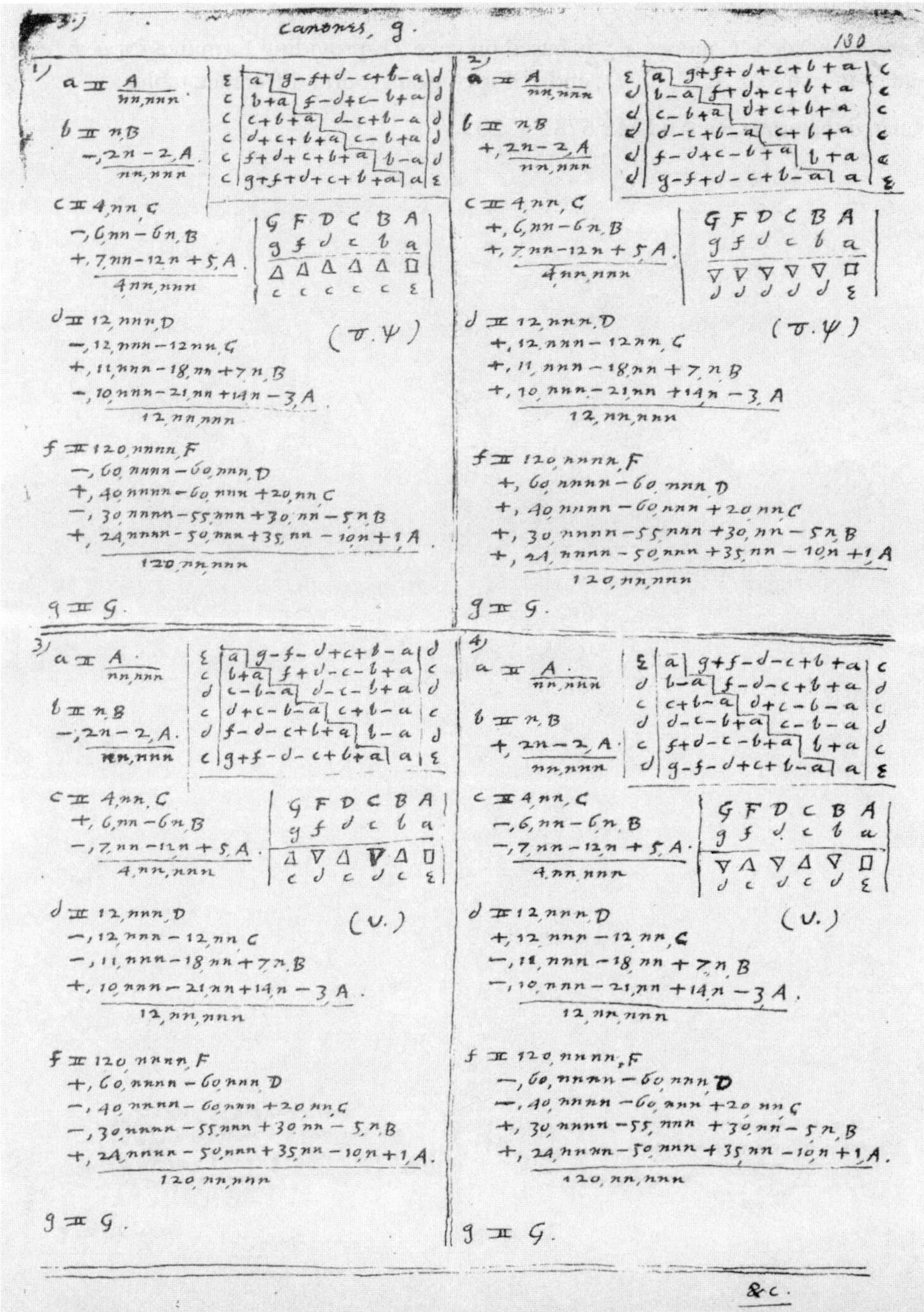
3.)
Canones, g.
130
1)
a ꞁꞁ A / nn,nnn
b ꞁꞁ nB
−,2n − 2,A / nn,nnn
ε | a | g−f+d−c+b−a | d
c | b+a | f−d+c−b+a | d
c | c+b+a | d−c+b−a | d
c | d+c+b+a | c−b+a | d
c | f+d+c+b+a | b−a | d
c | g+f+d+c+b+a | a | ε
c ꞁꞁ 4,nn,C
−,6nn−6n,B
+,7nn−12n+5,A / 4nn,nnn
G F D C B A
g f d c b a
Δ Δ Δ Δ Δ □
c c c c c ε
d ꞁꞁ 12,nnn,D
−,12,nnn−12nn,C
+,11,nnn−18,nn+7n,B
−,10,nnn−21,nn+14n−3,A / 12,nn,nnn
(ϖ.ψ)
f ꞁꞁ 120,nnnn,F
−,60,nnnn−60,nnn,D
+,40,nnnn−60,nnn+20,nn,C
−,30,nnnn−55,nnn+30,nn−5,n,B
+,24,nnnn−50,nnn+35,nn−10,n+1,A / 120,nn,nnn
g ꞁꞁ G.
2)
a ꞁꞁ A / nn,nnn
b ꞁꞁ n,B
+,2n−2,A / nn,nnn
ε | a | g+f+d+c+b+a | c
d | b−a | f+d+c+b+a | c
d | c−b+a | d+c+b+a | c
d | d−c+b−a | c+b+a | c
d | f−d+c−b+a | b+a | c
d | g−f+d−c+b−a | a | ε
c ꞁꞁ 4,nn,C
+,6,nn−6n,B
+,7,nn−12n+5,A / 4nn,nnn
G F D C B A
g f d c b a
∇ ∇ ∇ ∇ ∇ □
d d d d d ε
d ꞁꞁ 12,nnn,D
+,12,nnn−12nn,C
+,11,nnn−18,nn+7,n,B
+,10,nnn−21,nn+14n−3,A / 12,nn,nnn
(ϖ.ψ)
f ꞁꞁ 120,nnnn,F
+,60,nnnn−60,nnn,D
+,40,nnnn−60,nnn+20,nn,C
+,30,nnnn−55,nnn+30,nn−5,n,B
+,24,nnnn−50,nnn+35,nn−10,n+1,A / 120,nn,nnn
g ꞁꞁ G.
3)
a ꞁꞁ A / nn,nnn
b ꞁꞁ n,B
−,2n−2,A / nn,nnn
ε | a | g−f−d+c+b−a | d
c | b+a | f+d−c−b+a | c
d | c−b−a | d−c−b+a | d
c | d+c−b−a | c+b−a | c
d | f−d−c+b+a | b−a | d
c | g+f−d−c+b+a | a | ε
c ꞁꞁ 4,nn,C
+,6,nn−6n,B
−,7,nn−12n+5,A / 4,nn,nnn
G F D C B A
g f d c b a
Δ ∇ Δ ∇ Δ □
c d c d c ε
d ꞁꞁ 12,nnn,D
−,12,nnn−12,nn,C
−,11,nnn−18,nn+7,n,B
+,10,nnn−21,nn+14,n−3,A / 12,nn,nnn
(υ.)
f ꞁꞁ 120,nnnn,F
+,60,nnnn−60,nnn,D
−,40,nnnn−60,nnn+20,nn,C
−,30,nnnn−55,nnn+30,nn−5,n,B
+,24,nnnn−50,nnn+35,nn−10,n+1,A / 120,nn,nnn
g ꞁꞁ G.
4)
a ꞁꞁ A / nn,nnn
b ꞁꞁ n,B
+,2n−2,A / nn,nnn
ε | a | g+f−d−c+b+a | c
d | b−a | f−d−c+b+a | d
c | c+b−a | d+c−b−a | c
d | d−c−b+a | c−b−a | d
c | f+d−c−b+a | b+a | c
d | g−f−d+c+b−a | a | ε
c ꞁꞁ 4,nn,C
−,6,nn−6,n,B
−,7,nn−12n+5,A / 4,nn,nnn
G F D C B A
g f d c b a
∇ Δ ∇ Δ ∇ □
d c d c d ε
d ꞁꞁ 12,nnn,D
+,12,nnn−12,nn,C
−,11,nnn−18,nn+7,n,B
−,10,nnn−21,nn+14,n−3,A / 12,nn,nnn
(υ.)
f ꞁꞁ 120,nnnn,F
−,60,nnnn−60,nnn,D
−,40,nnnn−60,nnn+20,nn,C
+,30,nnnn−55,nnn+30,nn−5,n,B
+,24,nnnn−50,nnn+35,nn−10,n+1,A / 120,nn,nnn
g ꞁꞁ G.
&c.

Page 24 [Folio 131]

Page 24, headed 'Canones f', is based on page 21, providing formulae for a, b, c, d, and f in terms of A, B, C, D, and F for constant fourth difference tables.

Related material: BL Add MS 6782, f. 205.

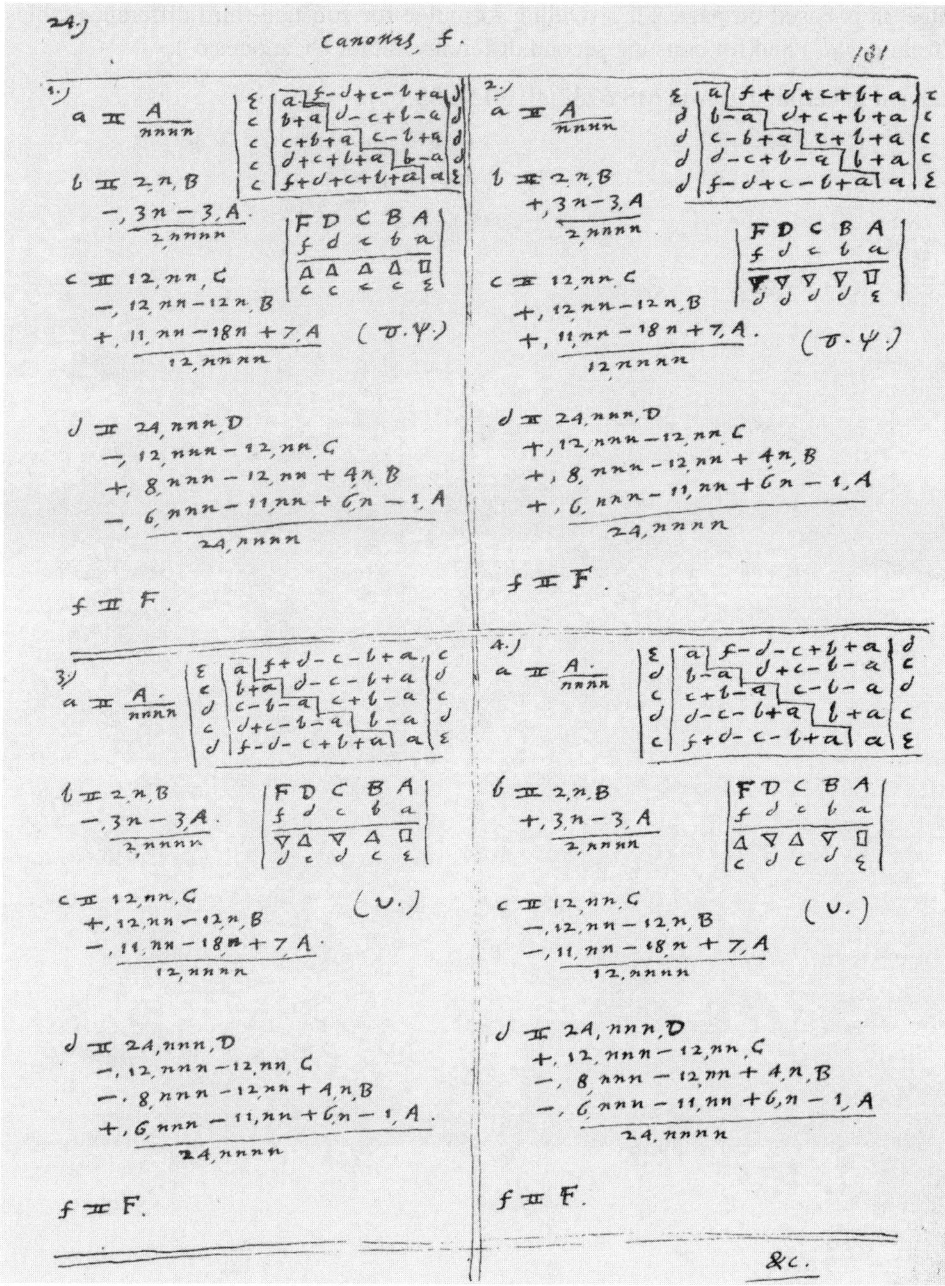

24.) Canones, f. 131

1.)

$a \sqcap \frac{A}{nnnn}$

ε	a	f−d+c−b+a	d
c	b+a	d−c+b−a	d
c	c+b+a	c−b+a	d
c	d+c+b+a	b−a	d
c	f+d+c+b+a	a	ε

$b \sqcap \frac{2,n,B \; -,3n-3,A}{2,nnnn}$

F	D	C	B	A
f	d	c	b	a
Δ	Δ	Δ	Δ	□
c	c	c	c	ε

$c \sqcap \frac{12,nn,C \; -,12,nn-12n,B \; +,11,nn-18n+7,A}{12,nnnn}$ (ϖ.ψ.)

$d \sqcap \frac{24,nnn,D \; -,12,nnn-12,nn,C \; +,8,nnn-12,nn+4n,B \; -,6,nnn-11,nn+6,n-1,A}{24,nnnn}$

$f \sqcap F.$

2.)

$a \sqcap \frac{A}{nnnn}$

ε	a	f+d+c+b+a	c
d	b−a	d+c+b+a	c
d	c−b+a	c+b+a	c
d	d−c+b−a	b+a	c
d	f−d+c−b+a	a	ε

$b \sqcap \frac{2,n,B \; +,3n-3,A}{2,nnnn}$

F	D	C	B	A
f	d	c	b	a
∇	∇	∇	∇	□
d	d	d	d	ε

$c \sqcap \frac{12,nn,C \; +,12,nn-12n,B \; +,11,nn-18n+7,A}{12nnnn}$ (ϖ.ψ.)

$d \sqcap \frac{24,nnn,D \; +,12,nnn-12,nn,C \; +,8,nnn-12,nn+4n,B \; +,6,nnn-11,nn+6n-1,A}{24,nnnn}$

$f \sqcap F.$

3.)

$a \sqcap \frac{A.}{nnnn}$

ε	a	f+d−c−b+a	c
c	b+a	d−c−b+a	d
d	c−b−a	c+b−a	c
c	d+c−b−a	b−a	d
d	f−d−c+b+a	a	ε

$b \sqcap \frac{2,n,B \; -,3n-3,A.}{2,nnnn}$

F	D	C	B	A
f	d	c	b	a
∇	Δ	∇	Δ	□
d	c	d	c	ε

$c \sqcap \frac{12,nn,C \; +,12,nn-12,n,B \; -,11,nn-18n+7,A}{12,nnnn}$ (ῡ.)

$d \sqcap \frac{24,nnn,D \; -,12,nnn-12,nn,C \; -,8,nnn-12,nn+4,n,B \; +,6,nnn-11,nn+6,n-1,A.}{24,nnnn}$

$f \sqcap F.$

4.)

$a \sqcap \frac{A.}{nnnn}$

ε	a	f−d−c+b+a	d
d	b−a	d+c−b−a	c
c	c+b−a	c−b−a	d
d	d−c−b+a	b+a	c
c	f+d−c−b+a	a	ε

$b \sqcap \frac{2,n,B \; +,3,n-3,A}{2,nnnn}$

F	D	C	B	A
f	d	c	b	a
Δ	∇	Δ	∇	□
c	d	c	d	ε

$c \sqcap \frac{12,nn,C \; -,12,nn-12,n,B \; -,11,nn-18,n+7,A}{12,nnnn}$ (ῡ.)

$d \sqcap \frac{24,nnn,D \; +,12,nnn-12,nn,C \; -,8,nnn-12,nn+4,n,B \; -,6,nnn-11,nn+6,n-1,A}{24,nnnn}$

$f \sqcap F.$

&c.

Page 25 [Folio 132]

Page 25 is based on page 22, providing formulae for constant third difference tables ('Canones *d*') and for constant second difference tables ('Canones *c*').

Related material: BL Add MS 6782, ff. 202–203.

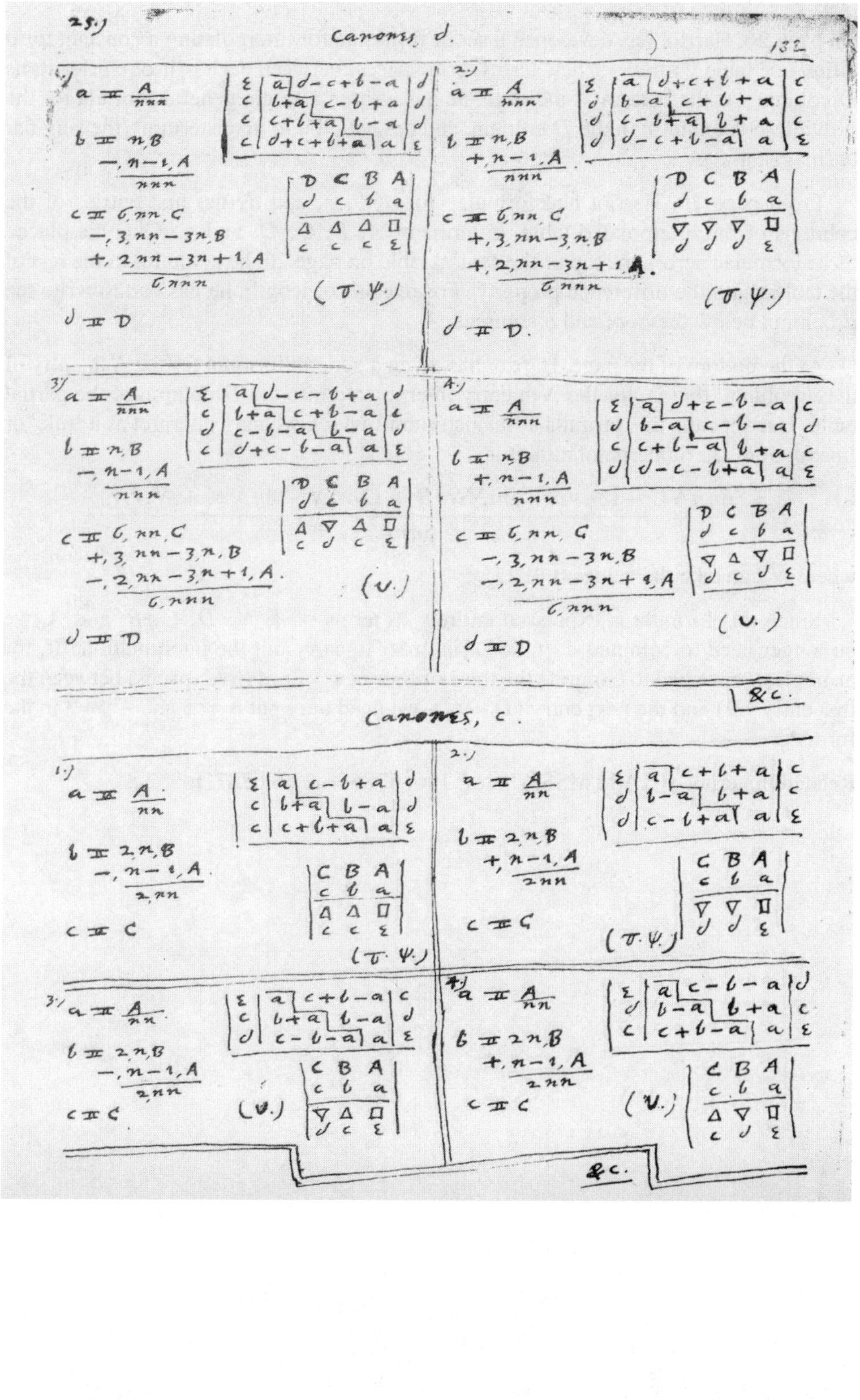

Page 26 [Folio 133]

On page 26, Harriot has developed a set of formulae for interpolating a constant third difference table with $n - 1$ new terms, or n spaces, between each pair of entries in its D column. At the bottom of the page, he has written a single general formula for the Nth interpolated term in the D column, and has called it a 'magisterium' (the singular of 'magisteria').

From page 25, Harriot had formulae for a, b, c, and d, the first entries of the columns of the interpolated table, in terms of A, B, C, D, and n. Once he placed these formulae across the top of the central table on page 26, he could fill in the rest of the table using the difference property. For reasons of length, he has had to write the d column below the a, b, and c columns.

At the bottom of the page, Harriot has given a general formula for the Nth entry of the d column; that is, for the Nth entry interpolated into the D column of the partial table. Harriot calls this formula a 'magisterium', which we may interpret as a 'rule' or 'instruction'. In modern notation it is

$$D + \frac{6nnNC - (3nnN - 3nNN)B + (2nnN - 3nNN + NNN)A}{6nnn},$$

where N can take the values $0, 1, 2, \ldots$.

Since this formula is expressed entirely in terms of n, N, D, C, B, and A, we no longer need to compute a, b, and c in order to carry out the interpolation. If, for example, we wished to compute the third of four new values (five spaces) between the first entry (D) and the next entry ($D + C$), we need only put $n = 5$ and $N = 3$ in the formula.

Related material: BL Add MSS 6782, f. 196; 6786, f. 227; 6787, f. 352v.

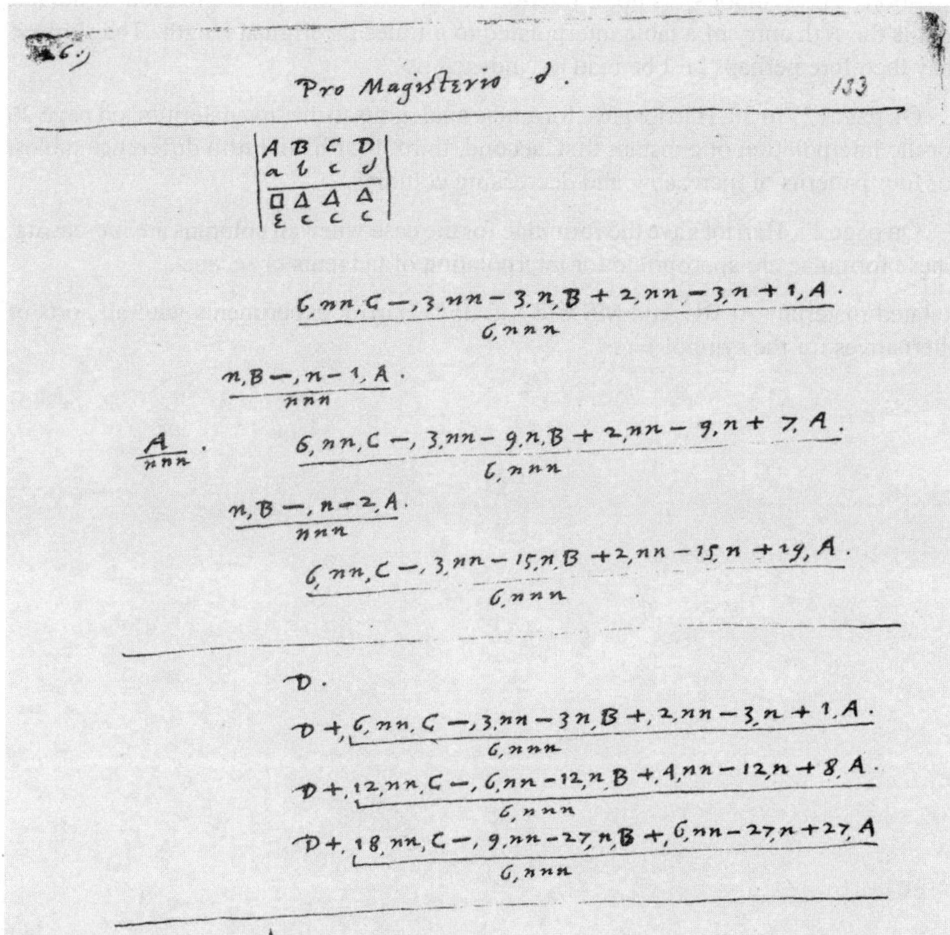

6.)

Pro Magisterio d. 133

A	B	C	D
a	b	c	d
□	Δ	Δ	Δ
ε	c	c	c

$$\frac{6,nn,C -,3,nn-3,n,B + 2,nn,-3,n+1,A.}{6,nnn}$$

$$\frac{n,B -,n-1,A.}{nnn}$$

$$\frac{A}{nnn}. \qquad \frac{6,nn,C -,3,nn-9,n,B + 2,nn-9,n+7,A.}{6,nnn}$$

$$\frac{n,B -,n-2,A.}{nnn}$$

$$\frac{6,nn,C -,3,nn-15,n,B + 2,nn-15,n+19,A.}{6,nnn}$$

D.

$$D + \frac{6,nn,C -,3,nn-3n,B +,2,nn-3,n+1,A.}{6,nnn}$$

$$D + \frac{12,nn,C -,6,nn-12,n,B +,4,nn-12,n+8,A.}{6,nnn}$$

$$D + \frac{18,nn,C -,9,nn-27,n,B +,6,nn-27,n+27,A}{6,nnn}$$

Magisterium. 7

$$D + \frac{6,\frac{nn}{x}\,C \;-,3,\frac{nn}{x} - \frac{3,n}{xx}\,B \;+,2,\frac{nn}{x} - \frac{3,n}{xx} + xxx,A.}{6,nnn.}$$

Page 27 [Folio 134]

Notation: The symbol $\frac{N}{n}$ is not a fraction but indicates that the expression to the left of it is the Nth entry of a table interpolated to n times its original length. The sign $)\!\!=\!\!($ may therefore perhaps best be read as 'indexed by'.

On pages 27 to 30, Harriot gave formulae analogous to the 'magisterium' on page 26 for the interpolation of constant first, second, third, fourth, and fifth difference tables, for four patterns of increasing and decreasing columns.

On page 27, Harriot gave the formulae for the case when all columns are increasing. These formulae are appropriate for interpolation of tangents or secants.

Related material: At BL Add MS 6787, f. 352, Harriot experiments with all sorts of alternatives for the symbol $)\!\!=\!\!($.

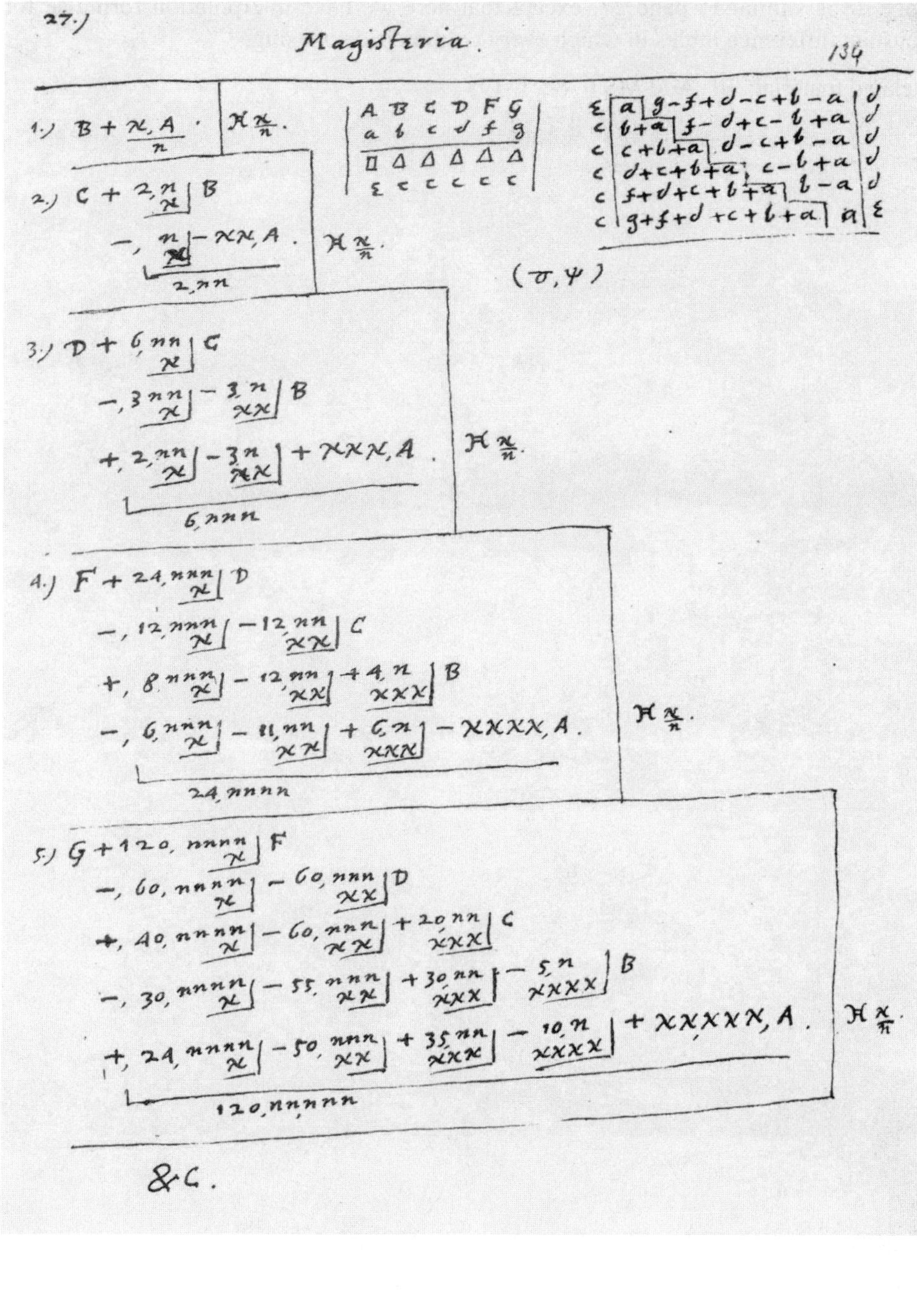
27.)

Magisteria.

134

1.) $B + \frac{x, A}{n}$. ♓ $\frac{x}{n}$.

2.) $C + \frac{2, n}{x} \mid B$

$-, \frac{n}{x} \mid - xx, A$. ♓ $\frac{x}{n}$.

$2, nn$

3.) $D + \frac{6, nn}{x} \mid C$

$-, \frac{3\,nn}{x} \mid - \frac{3\,n}{xx} \mid B$

$+, \frac{2, nn}{x} \mid - \frac{3n}{xx} \mid + xxx, A$. ♓ $\frac{x}{n}$.

$6, nnn$

4.) $F + \frac{24, nnn}{x} \mid D$

$-, \frac{12, nnn}{x} \mid - \frac{12\,nn}{xx} \mid C$

$+, \frac{8\,nnn}{x} \mid - \frac{12\,nn}{xx} \mid + \frac{4\,n}{xxx} \mid B$

$-, \frac{6\,nnn}{x} \mid - \frac{11, nn}{xx} \mid + \frac{6\,n}{xxx} \mid - xxxx, A$. ♓ $\frac{x}{n}$.

$24, nnnn$

5.) $G + \frac{120, nnnn}{x} \mid F$

$-, \frac{60, nnnn}{x} \mid - \frac{60, nnn}{xx} \mid D$

$+, \frac{40, nnnn}{x} \mid - \frac{60, nnn}{xx} \mid + \frac{20\,nn}{xxx} \mid C$

$-, \frac{30, nnnn}{x} \mid - \frac{55, nnn}{xx} \mid + \frac{30\,nn}{xxx} \mid - \frac{5, n}{xxxx} \mid B$

$+, \frac{24, nnnn}{x} \mid - \frac{50, nnn}{xx} \mid + \frac{35\,nn}{xxx} \mid - \frac{10, n}{xxxx} \mid + xxxxx, A$. ♓ $\frac{x}{n}$.

$120, nnnnn$

A	B	C	D	F	G
a	b	c	d	f	g
□	△	△	△	△	△
ε	c	c	c	c	c

ε	a | g − f + d − c + b − a	d
c	b + a | f − d + c − b + a	d
c	c + b + a | d − c + b − a	d
c	d + c + b + a | c − b + a	d
c	f + d + c + b + a | b − a	d
c	g + f + d + c + b + a | a	ε

(σ, ψ)

&c.

Page 28 [Folio 135]

Page 28 is similar to page 27, except that here we have interpolation formulae for constant difference tables in which every column is decreasing.

Related material: BL Add MS 6782, f. 193.

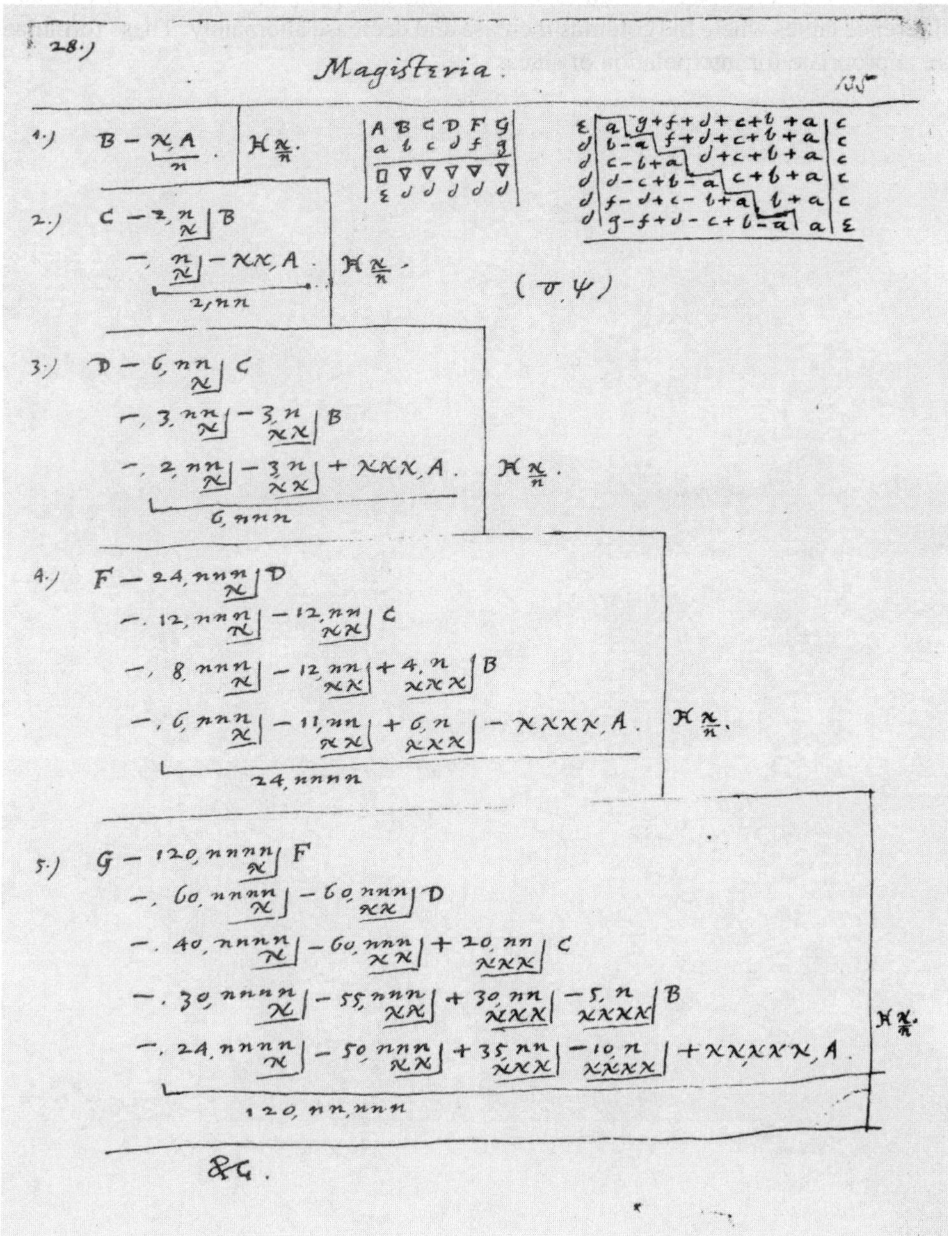

28.)

Magisteria. 135

A	B	C	D	F	G
a	b	c	d	f	g
□	▽	▽	▽	▽	▽
ε	d	d	d	d	d

ε | a | g + f + d + c + b + a | c
d | b − a | f + d + c + b + a | c
d | c − b + a | d + c + b + a | c
d | d − c + b − a | c + b + a | c
d | f − d + c − b + a | b + a | c
d | g − f + d − c + b − a | a | ε

(ʊ, ψ)

1.) B − x, A / n . ℋ x/n.

2.) C − 2, n/x | B
−, n/x | − xx, A . ℋ x/n.
2, nn

3.) D − 6, nn/x | C
−, 3, nn/x | − 3, n/xx | B
−, 2, nn/x | − 3 n/xx | + xxx, A . ℋ x/n.
6, nnn

4.) F − 24, nnn/x | D
−. 12, nnn/x | − 12, nn/xx | C
−, 8, nnn/x | − 12, nn/xx | + 4, n/xxx | B
−, 6, nnn/x | − 11, nn/xx | + 6, n/xxx | − xxxx, A . ℋ x/n.
24, nnnn

5.) G − 120, nnnn/x | F
−, 60, nnnn/x | − 60, nnn/xx | D
−. 40, nnnn/x | − 60, nnn/xx | + 20, nn/xxx | C
−. 30, nnnn/x | − 55, nnn/xx | + 30, nn/xxx | − 5, n/xxxx | B
−. 24, nnnn/x | − 50, nnn/xx | + 35, nn/xxx | − 10, n/xxxx | + xxxxx, A . ℋ x/n.
120, nn, nnn

&c.

Page 29 [Folio 136]

Page 29 is similar to pages 27 and 28; it contains interpolation formulae for constant difference tables where the columns increase and decrease alternately. These formulae are appropriate for interpolation of sines.

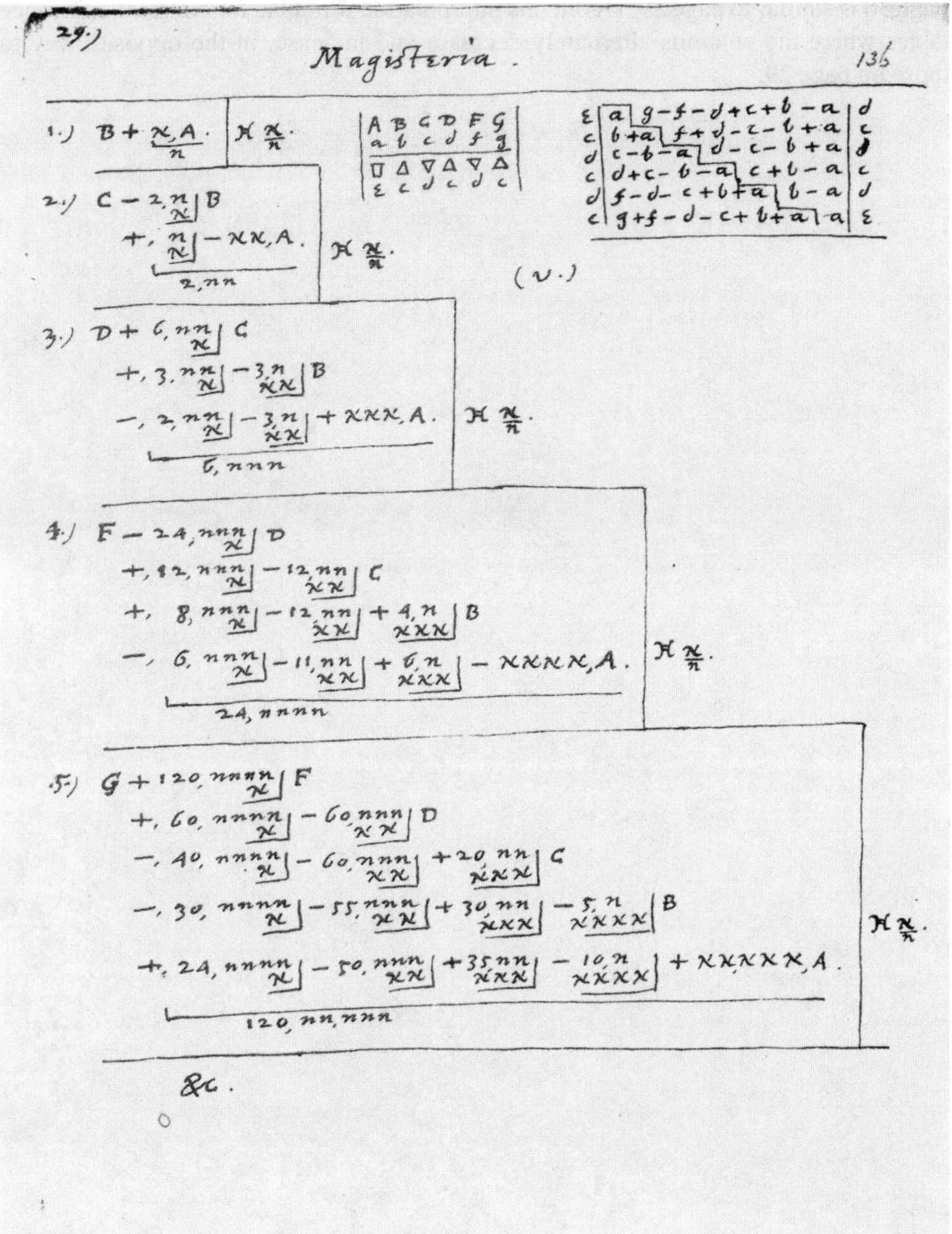

29.)

Magisteria. 136

1.) $B + \frac{x}{n}A.$ ℋ $\frac{x}{n}$.

2.) $C - 2, \frac{n}{x}\rfloor B$
$+, \frac{n}{x}\rfloor - xx, A.$ ℋ $\frac{x}{n}$.
$2, nn$

3.) $D + 6, \frac{nn}{x}\rfloor C$
$+, 3, \frac{nn}{x}\rfloor - \frac{3, n}{xx}\rfloor B$
$-, 2, \frac{nn}{x}\rfloor - \frac{3, n}{xx}\rfloor + xxx, A.$ ℋ $\frac{x}{n}$.
$6, nnn$

4.) $F - 24, \frac{nnn}{x}\rfloor D$
$+, 12, \frac{nnn}{x}\rfloor - \frac{12, nn}{xx}\rfloor C$
$+, 8, \frac{nnn}{x}\rfloor - \frac{12, nn}{xx}\rfloor + \frac{4, n}{xxx}\rfloor B$
$-, 6, \frac{nnn}{x}\rfloor - \frac{11, nn}{xx}\rfloor + \frac{6, n}{xxx}\rfloor - xxxx, A.$ ℋ $\frac{x}{n}$.
$24, nnnn$

5.) $G + 120, \frac{nnnn}{x}\rfloor F$
$+, 60, \frac{nnnn}{x}\rfloor - \frac{60, nnn}{xx}\rfloor D$
$-, 40, \frac{nnnn}{x}\rfloor - \frac{60, nnn}{xx}\rfloor + \frac{20, nn}{xxx}\rfloor C$
$-, 30, \frac{nnnn}{x}\rfloor - \frac{55, nnn}{xx}\rfloor + \frac{30, nn}{xxx}\rfloor - \frac{5, n}{xxxx}\rfloor B$
$+, 24, \frac{nnnn}{x}\rfloor - \frac{50, nnn}{xx}\rfloor + \frac{35, nn}{xxx}\rfloor - \frac{10, n}{xxxx}\rfloor + xxxxx, A.$ ℋ $\frac{x}{n}$.
$120, nn, nnn$

&c.

A	B	C	D	F	G
a	b	c	d	f	g
◻	△	▽	△	▽	△
ε	c	d	c	d	c

ε	a ⌉ g − f − d + c + b − a	d
c	b + a ⌉ f + d − c − b + a	c
d	c − b − a ⌉ d − c − b + a	d
c	d + c − b − a ⌉ c + b − a	c
d	f − d − c + b + a ⌉ b − a	d
c	g + f − d − c + b + a ⌉ a	ε

(v.)

Page 30 [Folio 137]

Page 30 is similar to page 29; it contains interpolation formulae for constant difference tables where the columns alternately decrease and increase, in the opposite way to those on page 29.

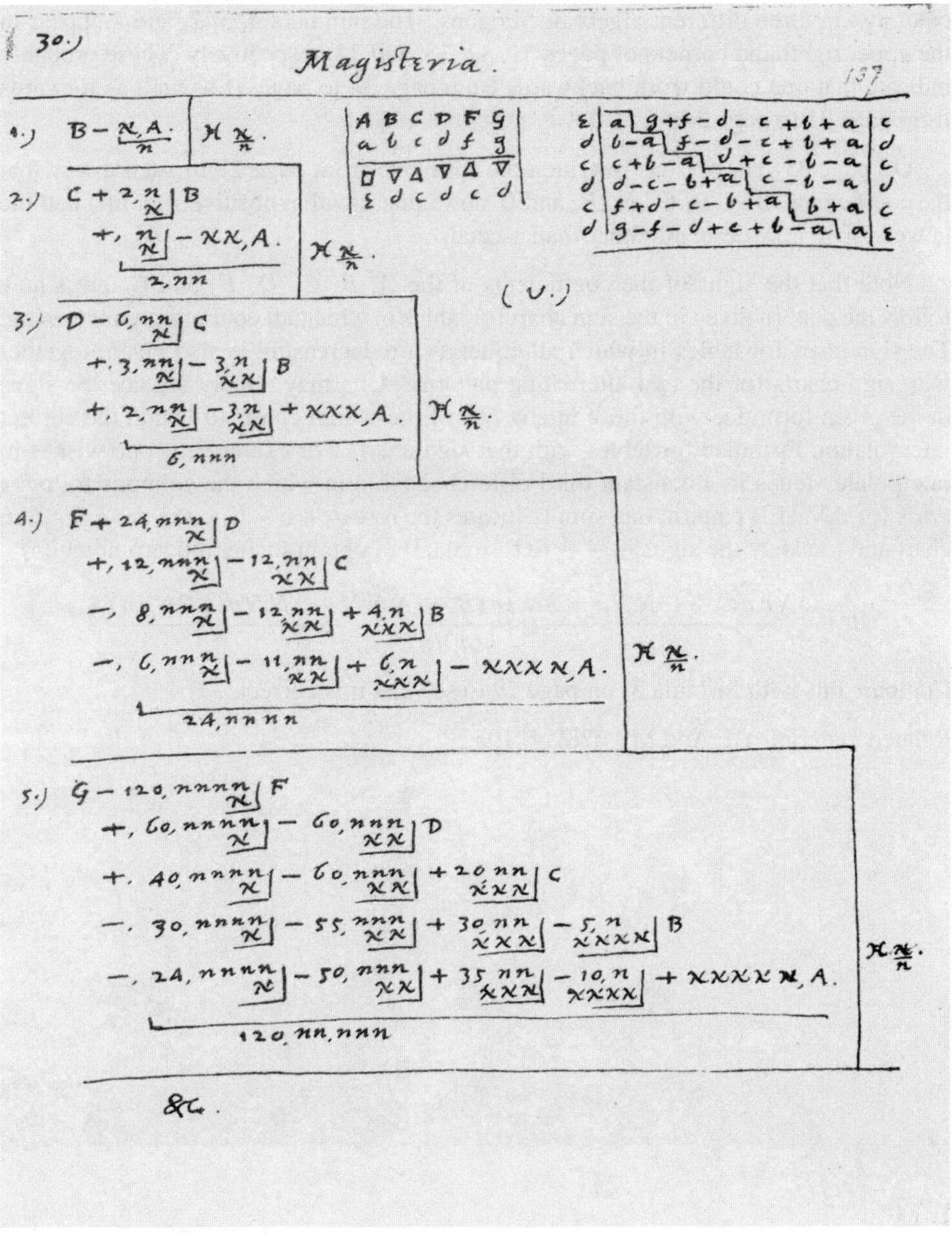
30.)
Magisteria.
137
1.) B − x,A / n | H x/n.
A B C D F G
a b c d f g
□ ▽ △ ▽ △ ▽
ε d c d c d
2.) C + 2,n/x B + ,n/x − xx,A. / 2,nn H x/n.
(V.)
3.) D − 6,nn/x C + 3,nn/x − 3,n/xx B + 2,nn/x − 3,n/xx + xxx,A. / 6,nnn H x/n.
4.) F + 24,nnn/x D + ,12,nnn/x − 12,nn/xx C −, 8,nnn/x − 12,nn/xx + 4,n/xxx B −, 6,nnn/x − 11,nn/xx + 6,n/xxx − xxxx,A. / 24,nnnn H x/n.
5.) G − 120,nnnn/x F +, 60,nnnn/x − 60,nnn/xx D +, 40,nnnn/x − 60,nnn/xx + 20,nn/xxx C −, 30,nnnn/x − 55,nnn/xx + 30,nn/xxx − 5,n/xxxx B −, 24,nnnn/x − 50,nnn/xx + 35,nn/xxx − 10,n/xxxx + xxxxx,A. / 120,nn,nnn H x/n.
&c.

Page 31 [Folio 138]

On pages 31 to 33, Harriot has rewritten the formulae from page 27 (for increasing columns) in three different algebraic versions. The numbers 4, 3, 2, and 1 appear in the upper righthand corners of pages 31, 32, 33, and 34, respectively. These probably indicate that one could work backwards from page 34 to page 31 as well as forwards from page 31 to page 34.

On page 31, Harriot has rewritten the formulae from page 27 in such a way that the coefficients of A, B, C, D, F, and G now appear with a positive sign first and the powers of N now descend rather than ascend.

Note that the signs of the coefficients of the A, B, C, D, F, and G terms now follow the pattern given in the sign chart for tables in which all columns are increasing. The sign chart for tables in which all columns are decreasing is also given, together with sign charts for the two alternating patterns. One may simply replace the signs in the given formulae with those in any one of these sign charts to obtain the correct interpolation formulae for tables with that sign chart. For example, if one wishes to interpolate values in a constant third difference table in which the columns follow a $cdc\varepsilon$ (or $\Delta\nabla\Delta\Box$) pattern, one simply locates the row $d + c - b - a$ in the $cdc\varepsilon$ sign chart and transfers the signs $+ - -$ to formula 3) to obtain (using modern notation)

$$D + \frac{6NnnC - (3NNn - 3Nnn)B - (NNN - 3NNn + 2Nnn)A}{6nnn}.$$

Compare this with formula 3) on page 29 to see that it is correct.

Related material: BL Add MS 6782, f. 195.

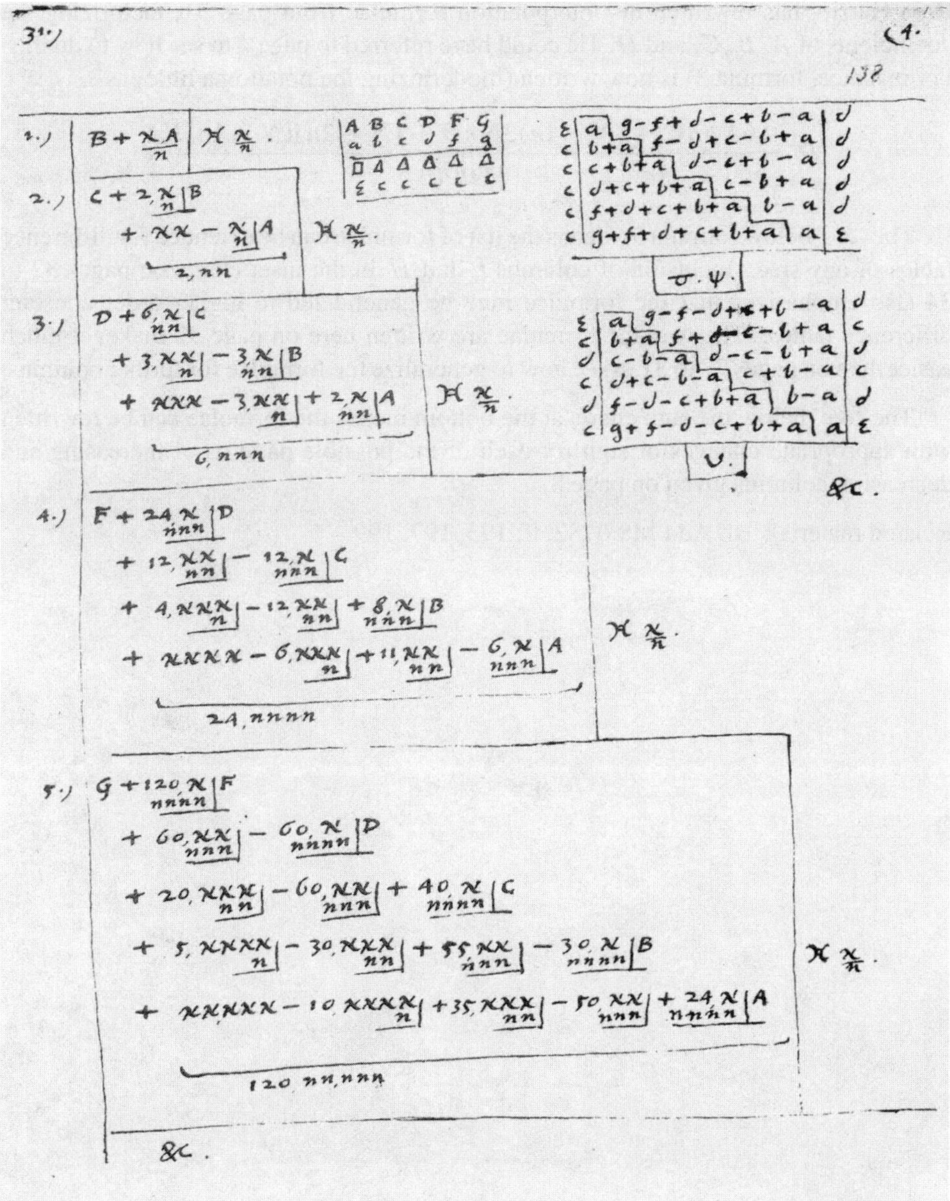
3'.)
(4.
138
1.) B + x/n |A ℋ x/n.
2.) C + 2,x/n |B
+ xx − x/n |A ℋ x/n.
2,nn
3.) D + 6,x/nn |C
+ 3,xx/n − 3,x/nn |B
+ xxx − 3,xx/n + 2,x/nn |A ℋ x/n.
6,nnn
4.) F + 24,x/nnn |D
+ 12,xx/nn − 12,x/nnn |C
+ 4,xxx/n − 12,xx/nn + 8,x/nnn |B
+ xxxx − 6,xxx/n + 11,xx/nn − 6,x/nnn |A ℋ x/n.
24,nnnn
5.) G + 120,x/nnnn |F
+ 60,xx/nnn − 60,x/nnnn |D
+ 20,xxx/nn − 60,xx/nnn + 40,x/nnnn |C
+ 5,xxxx/n − 30,xxx/nn + 55,xx/nnn − 30,x/nnnn |B
+ xxxxx − 10,xxxx/n + 35,xxx/nn − 50,xx/nnn + 24,x/nnnn |A ℋ x/n.
120,nnnnn
&c.
A B C D F G
a b c d f g
□ Δ Δ Δ Δ Δ
ε ε ε ε ε ε
ε a g − f + d − c + b − a d
ε b + a f − d + c − b + a d
ε c + b + a d − c + b − a d
ε d + c + b + a c − b + a d
ε f + d + c + b + a b − a d
ε g + f + d + c + b + a a ε
σ.ψ.
ε a g − f − d + c + b − a d
ε b + a f + d − c − b + a c
d c − b − a d − c − b + a d
c d + c − b − a c + b − a c
d f − d − c + b + a b − a d
c g + f − d − c + b + a a ε
υ.
&c.

Page 32 [Folio 139]

Here Harriot has rewritten his interpolation formulae from page 31, factorizing the coefficients of A, B, C, and D. He could have referred to page 4 to see how to do this. For instance, formula 3) is now written (modernizing the notation a little) as

$$D + \frac{6NnnC + (N - 1n)3NnB + (N - 2n)(N - 1n)NA}{6nnn}.$$

The '&c' below formula 5) means the list of formulae can be extended for difference tables of any size. Inclusion of columns h and H in the inset charts on pages 32 to 34 also emphasizes that the formulae may be generalized to higher order constant difference tables. The way the formulae are written here on page 32 makes it much easier than on pages 27 to 31 to see how to generalize the formulae for further columns.

The '&c' below the sign charts at the bottom means the formulae can be rewritten with appropriate changes of sign for each of the possible patterns of increasing and decreasing columns given on page 8.

Related material: BL Add MS 6782, ff. 193, 197, 199.

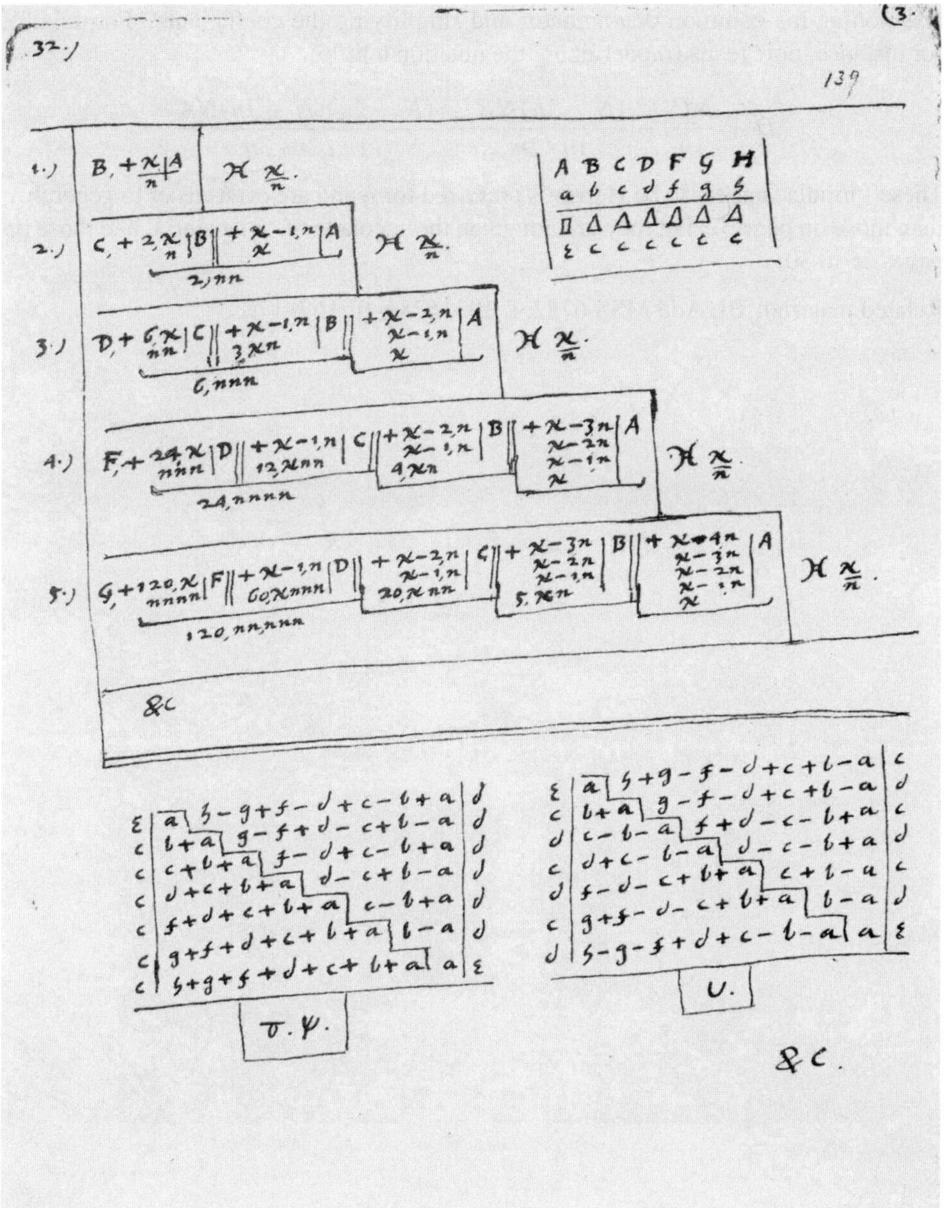

Page 33 [Folio 140]

On page 33 Harriot has rewritten the formulae from pages 31 and 32 yet again, this time abandoning his common denominator and simplifying the coefficients. Formula 3), for instance, now reads (modernizing the notation a little)

$$D + \frac{NC}{1n} + \frac{(N - 1n)NB}{1n \cdot 2n} + \frac{(N - 2n)(N - 1n)NA}{1n \cdot 2n \cdot 3n}.$$

These formulae appear to be Harriot's preferred form and are even easier to generalize than those on page 32, but they are not given the accolade of 'magisteria' like those on pages 26 to 30.

Related material: BL Add MSS 6782, f. 193; 6784, ff. 167–170.

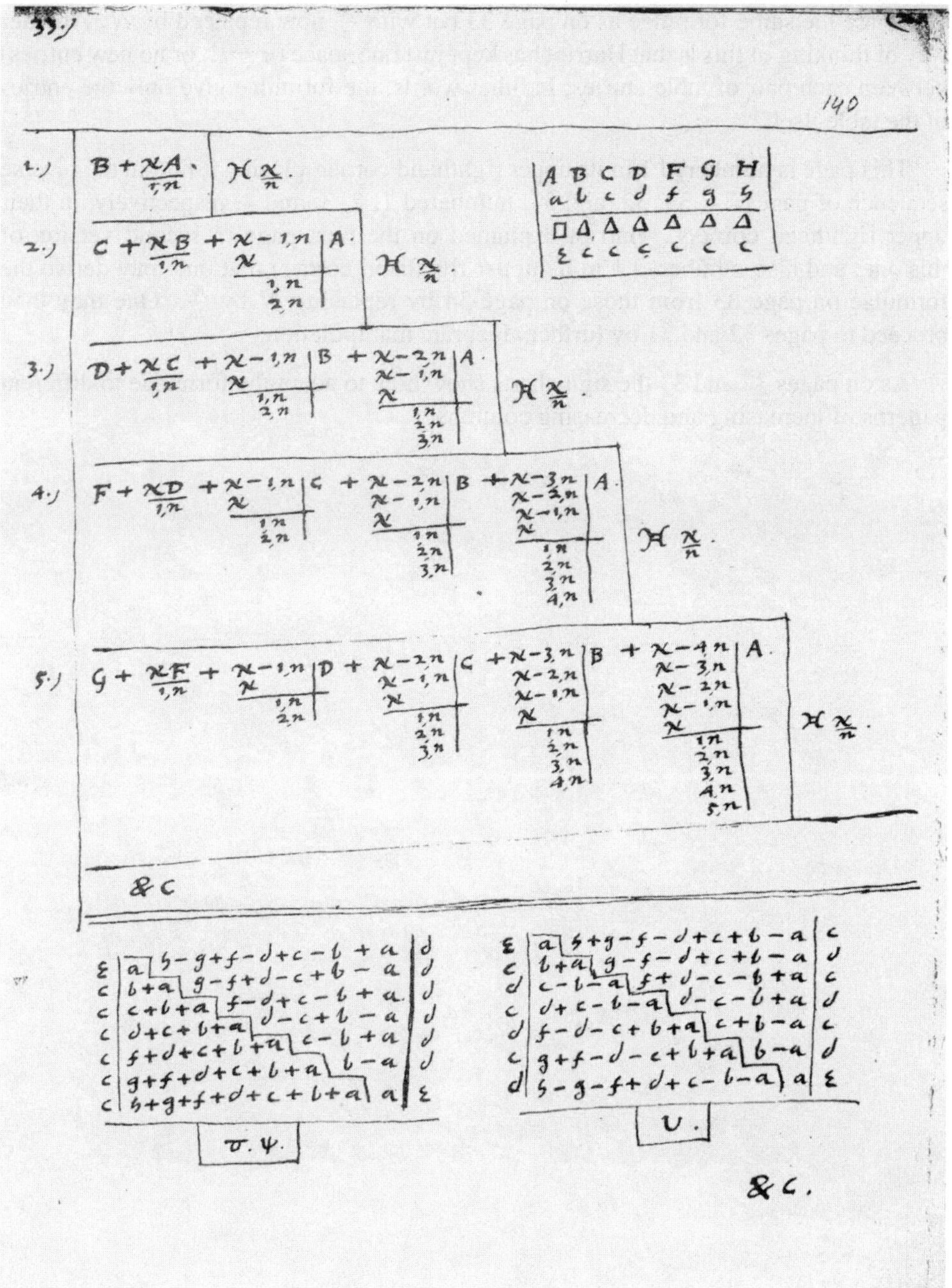

Page 34 [Folio 141]

These are the same formulae as on page 33 but with $\frac{N}{n}$ now replaced by N. Another way of thinking of this is that Harriot has kept just one space ($n = 1$, or no new entries) between each pair of table entries; in other words, the formulae give only the entries of the table itself.

This page is numbered 1 in its upper righthand corner, placing it first in the reverse sequence of pages 34, 33, 32, and 31, numbered 1, 2, 3, and 4, respectively, in their upper righthand corners. Harriot explained on the next page (a second version of this one, and also numbered 1 in its upper righthand corner) that one may derive the formulae on page 33 from those on page 34 by replacing N by $\frac{N}{n}$. One may then proceed to pages 32 and 31 by further algebraic manipulation.

As on pages 32 and 33 the sign charts show how to adapt the formulae to different patterns of increasing and decreasing columns.

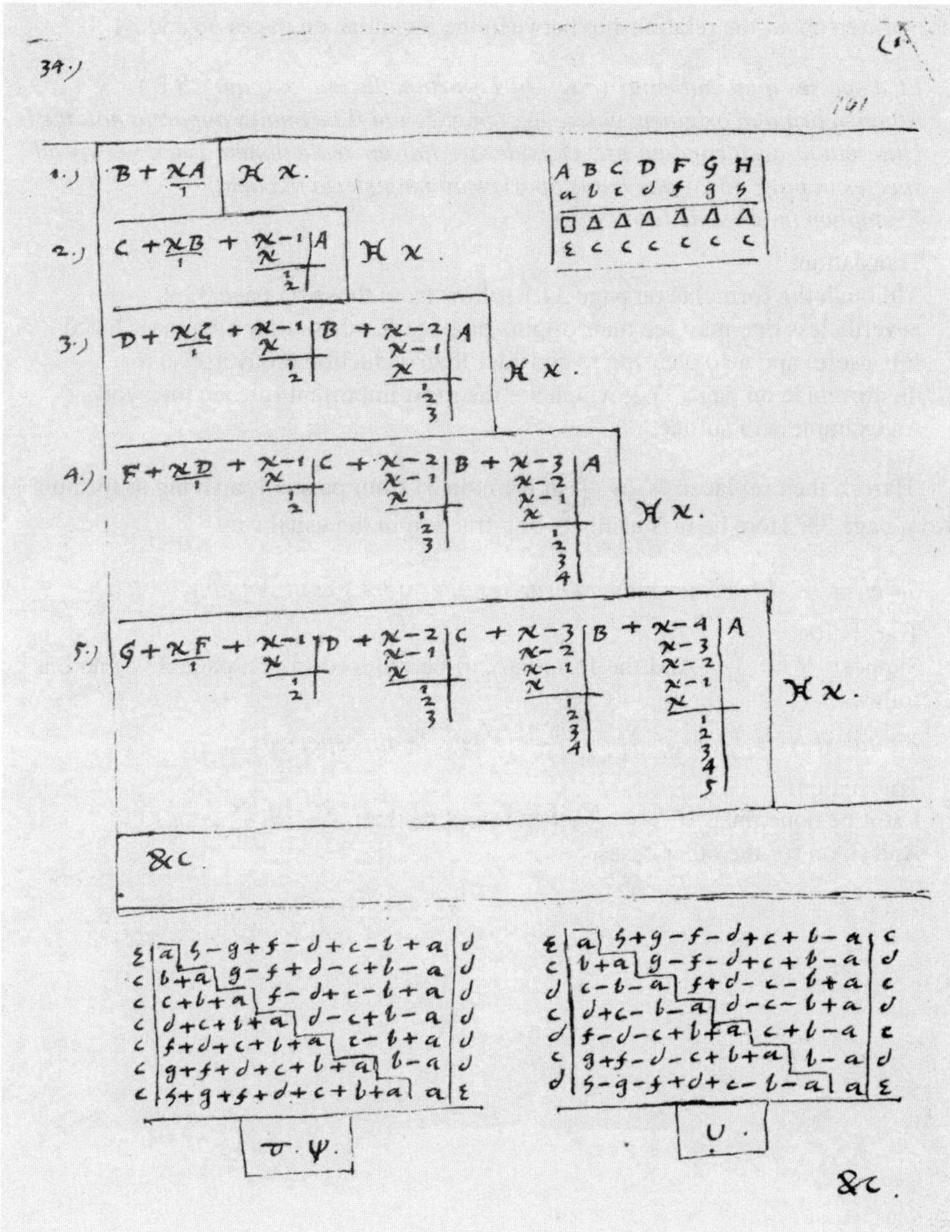

Page 34.2° [Folio 142]

Page 34.2° was probably added later (2° is an abbreviation for 'secundo'). Here Harriot has written about the relationship between the formulae on pages 33 and 34.

Et si species quae habentur (pag: 34.1.) ortum ducunt ex (pag: 33.2.)
Attamen primam originem videre licet pag. 5. ubi illae omnes apparent notatae.
Utile etiam ac incundum est, considerare harum reductionem (vice versa) ad species in pag: 33.2 quae huius operis sunt magisteria maxima.
Exemplum unum sufficiet.

Translation:
Although the formulae on page 34.1 follow from those on page 33.2,
nevertheless one may see their origins on page 5, where they all appear listed.
It is useful and also pleasing to consider their reduction (conversely) to
the formulae on page 33.2, which are the most important rules in this work.
An example will suffice.

Harriot then replaced N by $\frac{N}{n}$ in formula 3) from page 34, arriving at formula 3) from page 33. Here he *was* using $\frac{N}{n}$ as a fraction in the usual way.

Sit $N = \frac{N}{n}$. *Et species reducta erit: (ut pag: 33.2.) et ut sequitur:*

Translation:
Suppose $N = \frac{N}{n}$. And the formula will be reduced (as on page 33.2) and as follows:

Fit ita: si $N = \frac{N}{n}$, *erit:* $\frac{NC}{1} = \frac{NC}{1n}$. *Et sic de aliis speciebus.*

Translation:
Let it be done thus: If $N = \frac{N}{n}$, then it will be that: $\frac{NC}{1} = \frac{NC}{1n}$.
And so on for the other cases.

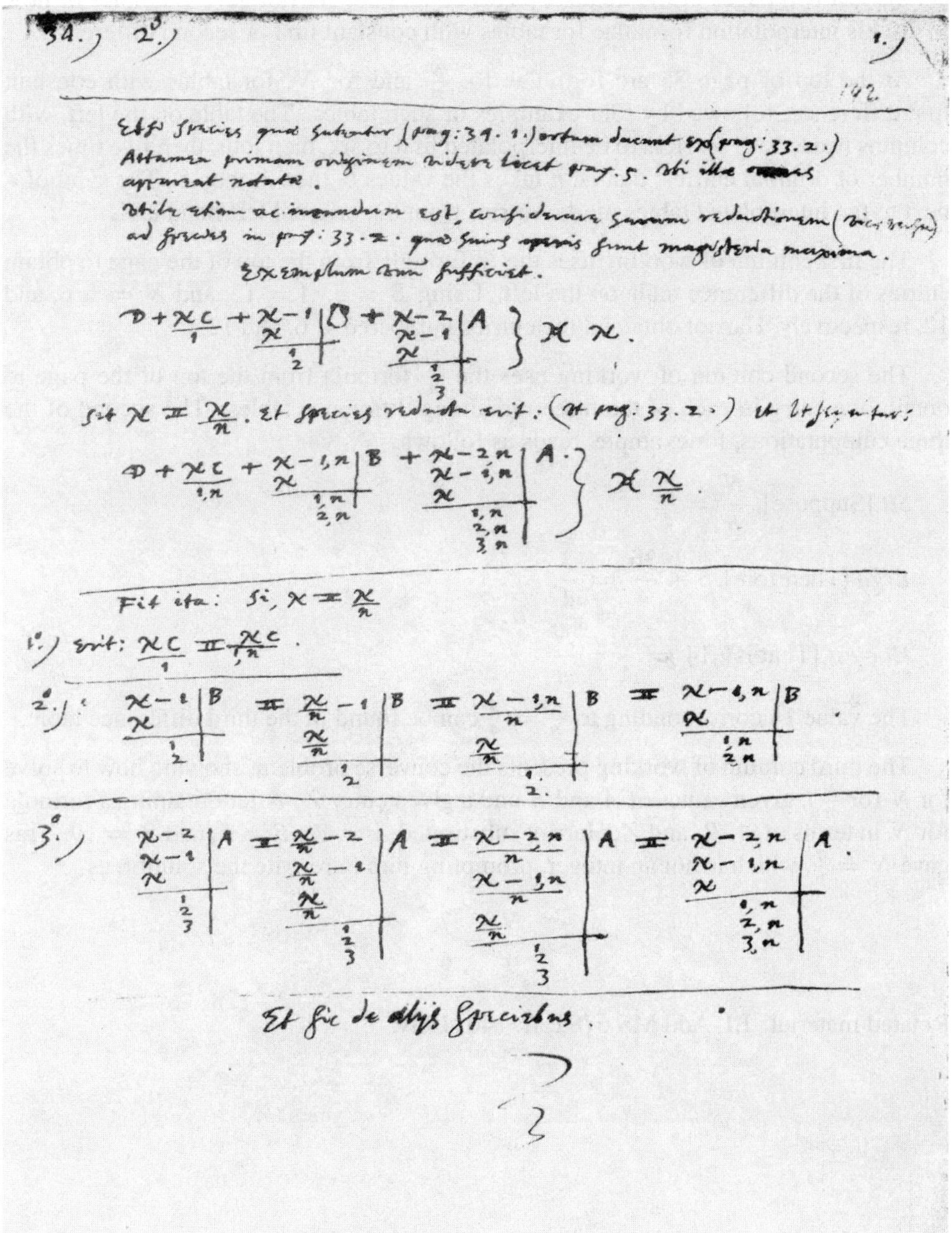

34.) 2°.) 1.)

142.

Istae species quae sequuntur (pag: 39. 1.) ortum ducunt ex (pag: 33. 2.)
Attamen primam originem videre licet pag. 5. ubi illae [illegible] appareant notatae.
Utile etiam ac iucundum est, considerare harum reductionem (vice versa) ad species in pag: 33. 2. quod huius operis sunt magisteria maxima.
Exemplum unum sufficiet.

$$D + \frac{\mathcal{X}C}{1} + \frac{\mathcal{X}-1 \cdot \mathcal{X}}{1 \cdot 2} B + \frac{\mathcal{X}-2 \cdot \mathcal{X}-1 \cdot \mathcal{X}}{1 \cdot 2 \cdot 3} A \Big\} \mathcal{X}\ x.$$

Sit $\mathcal{X} = \frac{\mathcal{X}}{n}$. Et species reducta erit: (ut pag. 33. 2.) ut sequitur:

$$D + \frac{\mathcal{X}C}{1,n} + \frac{\mathcal{X}-1,n \cdot \mathcal{X}}{1,n \cdot 2,n} B + \frac{\mathcal{X}-2,n \cdot \mathcal{X}-1,n \cdot \mathcal{X}}{1,n \cdot 2,n \cdot 3,n} A \Big\} \mathcal{X}\ \frac{\mathcal{X}}{n}.$$

Fit ita: Si, $\mathcal{X} = \frac{\mathcal{X}}{n}$

1°.) erit: $\frac{\mathcal{X}C}{1} = \frac{\mathcal{X}C}{1,n}$.

2°.) $\frac{\mathcal{X}-1 \cdot \mathcal{X}}{1 \cdot 2} B = \frac{\frac{\mathcal{X}}{n}-1 \cdot \frac{\mathcal{X}}{n}}{1 \cdot 2} B = \frac{\frac{\mathcal{X}-1,n}{n} \cdot \frac{\mathcal{X}}{n}}{1 \cdot 2} B = \frac{\mathcal{X}-1,n \cdot \mathcal{X}}{1,n \cdot 2,n} B$

3°.) $\frac{\mathcal{X}-2 \cdot \mathcal{X}-1 \cdot \mathcal{X}}{1 \cdot 2 \cdot 3} A = \frac{\frac{\mathcal{X}}{n}-2 \cdot \frac{\mathcal{X}}{n}-1 \cdot \frac{\mathcal{X}}{n}}{1 \cdot 2 \cdot 3} A = \frac{\frac{\mathcal{X}-2,n}{n} \cdot \frac{\mathcal{X}-1,n}{n} \cdot \frac{\mathcal{X}}{n}}{1 \cdot 2 \cdot 3} A = \frac{\mathcal{X}-2,n \cdot \mathcal{X}-1,n \cdot \mathcal{X}}{1,n \cdot 2,n \cdot 3,n} A.$

Et sic de aliis Speciebus

Page 35 [Folio 143]

On the last two pages of the treatise, pages 35 and 36, Harriot gave examples of how to use his interpolation formulae for tables with constant first or second difference.

At the top of page 35 are formulae for $\frac{N}{n}$ and for N, for tables with constant first difference, followed by four examples of such tables. The table on the left, with columns headed N, B, A, is to be interpolated first to six, then four, then five times the number of original entries; that is, n takes the values 6, then 4, then 5. The symbol $*$ next to the interpolated tables marks entries from the original left-hand table.

The first column of working uses the N formula from the top of the page to obtain entries of the difference table on the left. Using $B = 5$, $A = 12$, and $N = 1, 6$, and 12, respectively, Harriot obtained the entries numbered 1, 6, and 12.

The second column of working uses the $\frac{N}{n}$ formula from the top of the page to obtain one entry in each of the three remaining difference tables. The second of the three computations, for example, reads as follows.

Sit [Suppose], $\dfrac{N}{n} = \dfrac{3}{4}$.

Ergo [Therefore], $5 + \dfrac{36}{4} \,)\!\!=\!\!(\, \dfrac{3}{4}$.

Hoc est [That is], $14 \,)\!\!=\!\!(\, \dfrac{3}{4} = \dfrac{N}{n}$.

The value 14 corresponding to $\frac{N}{n} = \frac{3}{4}$ can be found in the third difference table.

The third column of working presents the converse problem, showing how to solve for N (or $\frac{N}{n}$), given values of A and B and a given entry Z. After obtaining a formula for N in terms of A, B, and Z, Harriot substituted $A = 12$, $B = 5$, and $Z = 19$. This gave $N = \frac{7}{6}$, which is not an integer, prompting him to rewrite the solution as

$$\frac{N}{n} = \frac{7}{6}.$$

Related material: BL Add MS 6782, ff. 146, 193v.

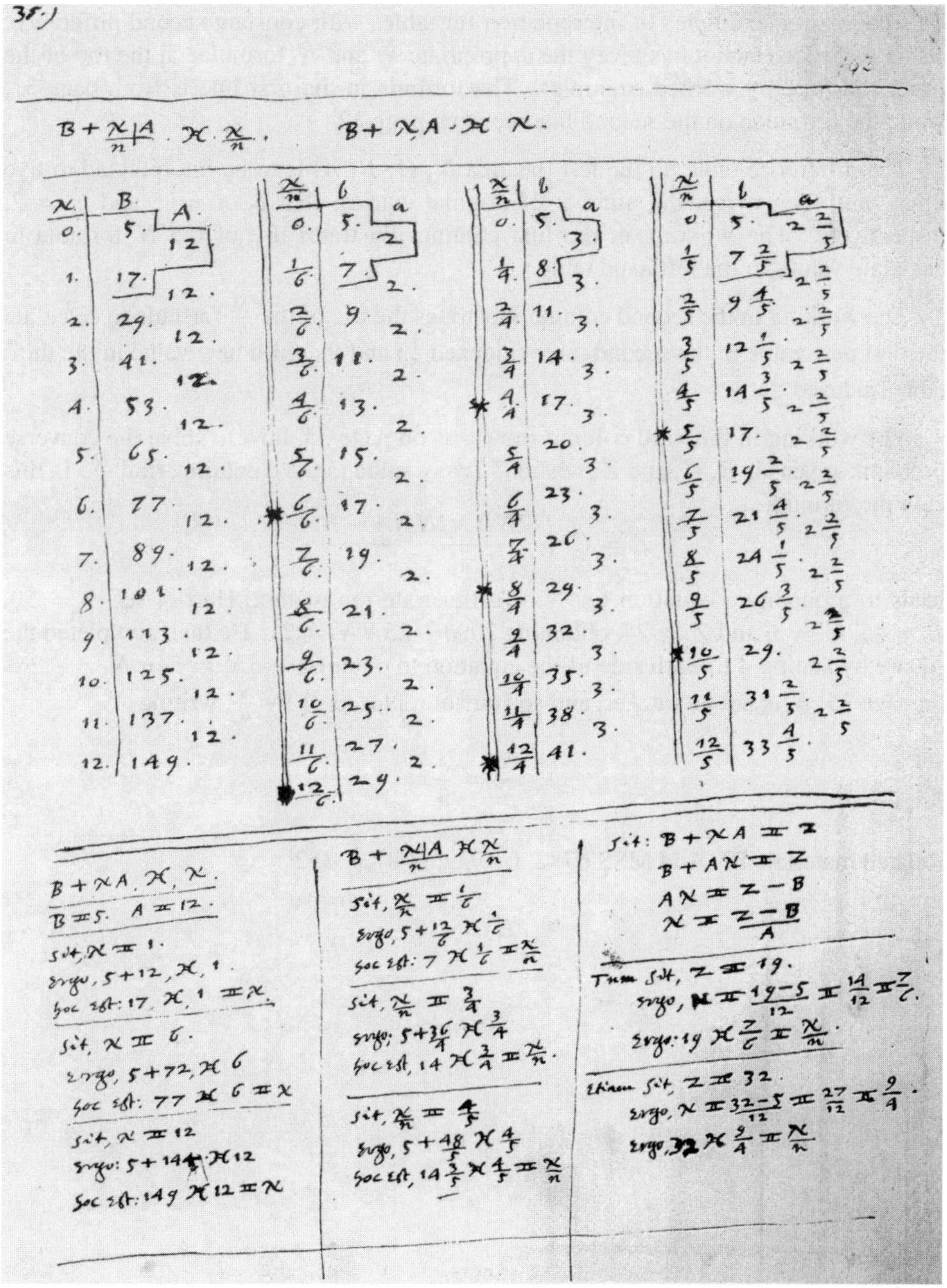

35.)

$B + \frac{x}{n}A \;\mathcal{H}\; \frac{x}{n}$. $B + xA \;\mathcal{H}\; x$.

x	B	A
0.	5.	
		12
1.	17.	
		12.
2.	29.	
		12.
3.	41.	
		12.
4.	53.	
		12.
5.	65.	
		12.
6.	77.	
		12.
7.	89.	
		12.
8.	101.	
		12.
9.	113.	
		12.
10.	125.	
		12.
11.	137.	
		12.
12.	149.	

$\frac{x}{n}$	b	a
0.	5.	
		2
$\frac{1}{6}$	7.	
		2.
$\frac{2}{6}$	9.	
		2.
$\frac{3}{6}$	11.	
		2.
$\frac{4}{6}$	13.	
		2.
$\frac{5}{6}$	15.	
		2.
* $\frac{6}{6}$	17.	
		2.
$\frac{7}{6}$	19.	
		2.
$\frac{8}{6}$	21.	
		2.
$\frac{9}{6}$	23.	
		2.
$\frac{10}{6}$	25.	
		2.
$\frac{11}{6}$	27.	
		2.
* $\frac{12}{6}$	29.	

$\frac{x}{n}$	b	a
0.	5.	
		3.
$\frac{1}{4}$	8.	
		3.
$\frac{2}{4}$	11.	
		3.
$\frac{3}{4}$	14.	
		3.
* $\frac{4}{4}$	17.	
		3.
$\frac{5}{4}$	20.	
		3.
$\frac{6}{4}$	23.	
		3.
$\frac{7}{4}$	26.	
		3.
* $\frac{8}{4}$	29.	
		3.
$\frac{9}{4}$	32.	
		3.
$\frac{10}{4}$	35.	
		3.
$\frac{11}{4}$	38.	
		3.
* $\frac{12}{4}$	41.	

$\frac{x}{n}$	b	a
0.	5.	
		$2\frac{2}{5}$
$\frac{1}{5}$	$7\frac{2}{5}$	
		$2\frac{2}{5}$
$\frac{2}{5}$	$9\frac{4}{5}$	
		$2\frac{2}{5}$
$\frac{3}{5}$	$12\frac{1}{5}$	
		$2\frac{2}{5}$
$\frac{4}{5}$	$14\frac{3}{5}$	
		$2\frac{2}{5}$
* $\frac{5}{5}$	17.	
		$2\frac{2}{5}$
$\frac{6}{5}$	$19\frac{2}{5}$	
		$2\frac{2}{5}$
$\frac{7}{5}$	$21\frac{4}{5}$	
		$2\frac{2}{5}$
$\frac{8}{5}$	$24\frac{1}{5}$	
		$2\frac{2}{5}$
$\frac{9}{5}$	$26\frac{3}{5}$	
		$2\frac{2}{5}$
* $\frac{10}{5}$	29.	
		$2\frac{2}{5}$
$\frac{11}{5}$	$31\frac{2}{5}$	
		$2\frac{2}{5}$
$\frac{12}{5}$	$33\frac{4}{5}$	

$B + xA.\ \mathcal{H},\ x.$
$B = 5.\quad A = 12.$
Sit, $x = 1$.
ergo, $5+12,\ \mathcal{H},\ 1$.
hoc est: $17.\ \mathcal{H},\ 1 = x$.
Sit, $x = 6$.
ergo, $5+72,\ \mathcal{H}\ 6$
hoc est: $77\ \mathcal{H}\ 6 = x$
Sit, $x = 12$
ergo: $5+144\ \mathcal{H}\ 12$
hoc est: $149\ \mathcal{H}\ 12 = x$

$B + \frac{x}{n}A\ \mathcal{H}\ \frac{x}{n}$
Sit, $\frac{x}{n} = \frac{1}{6}$
ergo, $5 + \frac{12}{6}\ \mathcal{H}\ \frac{1}{6}$
hoc est: $7\ \mathcal{H}\ \frac{1}{6} = \frac{x}{n}$
Sit, $\frac{x}{n} = \frac{3}{4}$
ergo; $5+\frac{36}{4}\ \mathcal{H}\ \frac{3}{4}$
hoc est, $14\ \mathcal{H}\ \frac{3}{4} = \frac{x}{n}$.
Sit, $\frac{x}{n} = \frac{4}{5}$
ergo, $5+\frac{48}{5}\ \mathcal{H}\ \frac{4}{5}$
hoc est, $14\frac{3}{5}\ \mathcal{H}\ \frac{4}{5} = \frac{x}{n}$

Sit: $B + xA = Z$
$B + Ax = Z$
$Ax = Z - B$
$x = \frac{Z-B}{A}$
Tum sit, $Z = 19$.
ergo, $x = \frac{19-5}{12} = \frac{14}{12} = \frac{7}{6}$.
Ergo: $19\ \mathcal{H}\ \frac{7}{6} = \frac{x}{n}$.
Etiam sit, $Z = 32$.
ergo, $x = \frac{32-5}{12} = \frac{27}{12} = \frac{9}{4}$.
ergo, $32\ \mathcal{H}\ \frac{9}{4} = \frac{x}{n}$

Page 36 [Folio 144]

This page gives examples of interpolation for tables with constant second difference. As on page 35, Harriot has given the appropriate $\frac{N}{n}$ and N formulae at the top of the page, followed by worked examples. The formula in the first line is from page 33, while the formulae on the second line are from page 32.

The difference table on the left (headed N, C, B, A) is to be interpolated to five times, and then twice, the number of original entries; that is, $n = 5$ and $n = 2$, respectively. The working in the first column illustrates use of the N formula to calculate values in the left hand table.

The working in the second column illustrates the use of the $\frac{N}{n}$ formula to calculate the first new value in the second table (indexed $\frac{1}{5}$) and the third new value in the third table (indexed $\frac{3}{2}$).

The working in the third column shows, as on page 35, how to solve the converse problem: given A, B, C, and Z, where Z is any value in the c column, find $\frac{N}{n}$. In this case the formula

$$Z = C + \frac{2NB + NNA - NA}{2}$$

leads to a quadratic equation for N. To illustrate the method Harriot set $A = 50$, $B = 45$, $C = 3$, and $Z = 24$, obtaining $20N + 25NN = 21$. He then completed the square by adding 4 to each side of the equation to obtain $2 + 5N = 5$ or $N = \frac{3}{5}$. As on page 35, N is not an integer, and so Harriot replaced N by $\frac{N}{n}$, writing

$$\frac{N}{n} = \frac{3}{5}.$$

Related material: BL Add MSS 6782, f. 193v; 6787, f. 352.

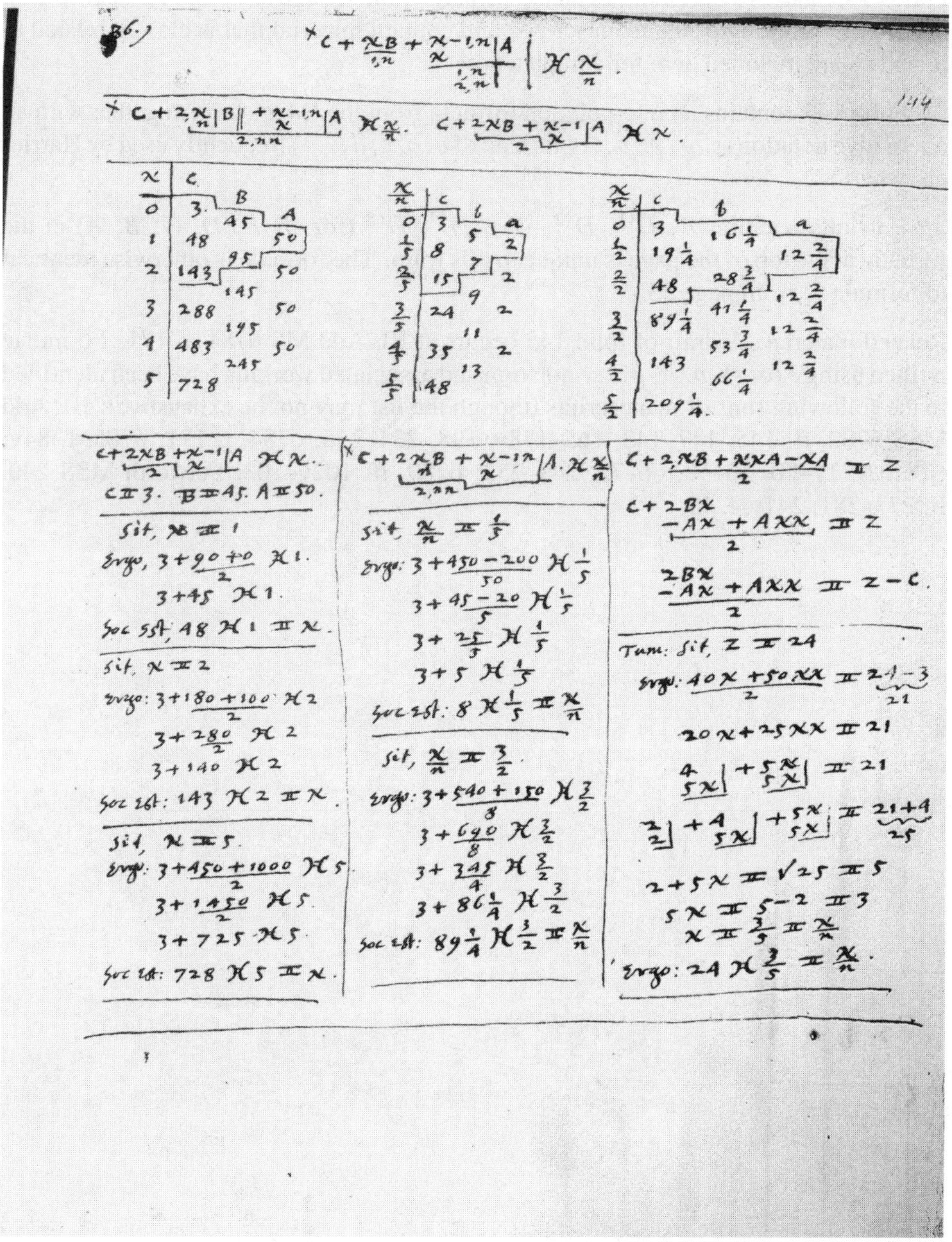
36.)
144
x	C	B	A
0	3	45	50
1	48	95	50
2	143	145	50
3	288	195	50
4	483	245	50
5	728		
Sit, x ꝏ 1
Ergo, 3 + 90 + 0 / 2 ⊁ 1.
3 + 45 ⊁ 1.
Hoc est, 48 ⊁ 1 ꝏ x.
Sit, x ꝏ 2
Ergo: 3 + 180 + 100 / 2 ⊁ 2
3 + 280 / 2 ⊁ 2
3 + 140 ⊁ 2
Hoc est: 143 ⊁ 2 ꝏ x.
Sit, x ꝏ 5.
Ergo: 3 + 450 + 1000 / 2 ⊁ 5.
3 + 1450 / 2 ⊁ 5.
3 + 725 ⊁ 5
Hoc est: 728 ⊁ 5 ꝏ x.
Sit, x/n ꝏ 1/5
Ergo: 3 + 450 − 200 / 50 ⊁ 1/5
3 + 45 − 20 / 5 ⊁ 1/5
3 + 25/5 ⊁ 1/5
3 + 5 ⊁ 1/5
Hoc est: 8 ⊁ 1/5 ꝏ x/n
Sit, x/n ꝏ 3/2.
Ergo: 3 + 540 + 150 / 8 ⊁ 3/2
3 + 690/8 ⊁ 3/2
3 + 345/4 ⊁ 3/2
3 + 86 1/4 ⊁ 3/2
Hoc est: 89 1/4 ⊁ 3/2 ꝏ x/n
Tum: Sit, Z ꝏ 24
20x + 25xx ꝏ 21
2 + 5x ꝏ √25 ꝏ 5
5x ꝏ 5 − 2 ꝏ 3
x ꝏ 3/5 ꝏ x/n
Ergo: 24 ⊁ 3/5 ꝏ x/n.

Endmatter [Folio 145]

Folios 145 and 146v are not paginated as part of the 'Magisteria' but lie next to it in the present ordering of the manuscripts, and contain material that is closely related to it, and so are included here for completeness.

Folio 145 contains an interpolation formula from the 'Magisteria' together with an alternative notation (ε, $\overset{1}{p}$, $\overset{2}{p}$, $\overset{3}{p}, \dots$ instead of $a, b, c, d, \dots$) frequently used by Harriot elsewhere.

Harriot's notation P, $D^{(1}$, $D^{(2}$, $D^{(3}$, $D^{(4}$, $D^{(5}$ (for G, F, D, C, B, A) in the formula at the top of the page is unique to this folio. The formula is otherwise identical to formula 5) from page 33.

Related material: A draft of folio 145 occurs at BL Add MS 6784, f. 171. Formulae written using e (or ε), p, $\overset{2}{p}$, $\overset{3}{p}, \dots$ notation and associated working have been identified in the following runs of manuscripts (though the list may not be exhaustive): BL Add MSS 6782, ff. 145, 147–148, 165–178v, 198, 234–236; 6784, f. 171; 6785, f. 84v; 6787, ff. 17–20, 53–58, 66–73, 249–252; 6789, ff. 102v–103; Petworth MSS 240, ff. 277–281; 241v, f. 11.

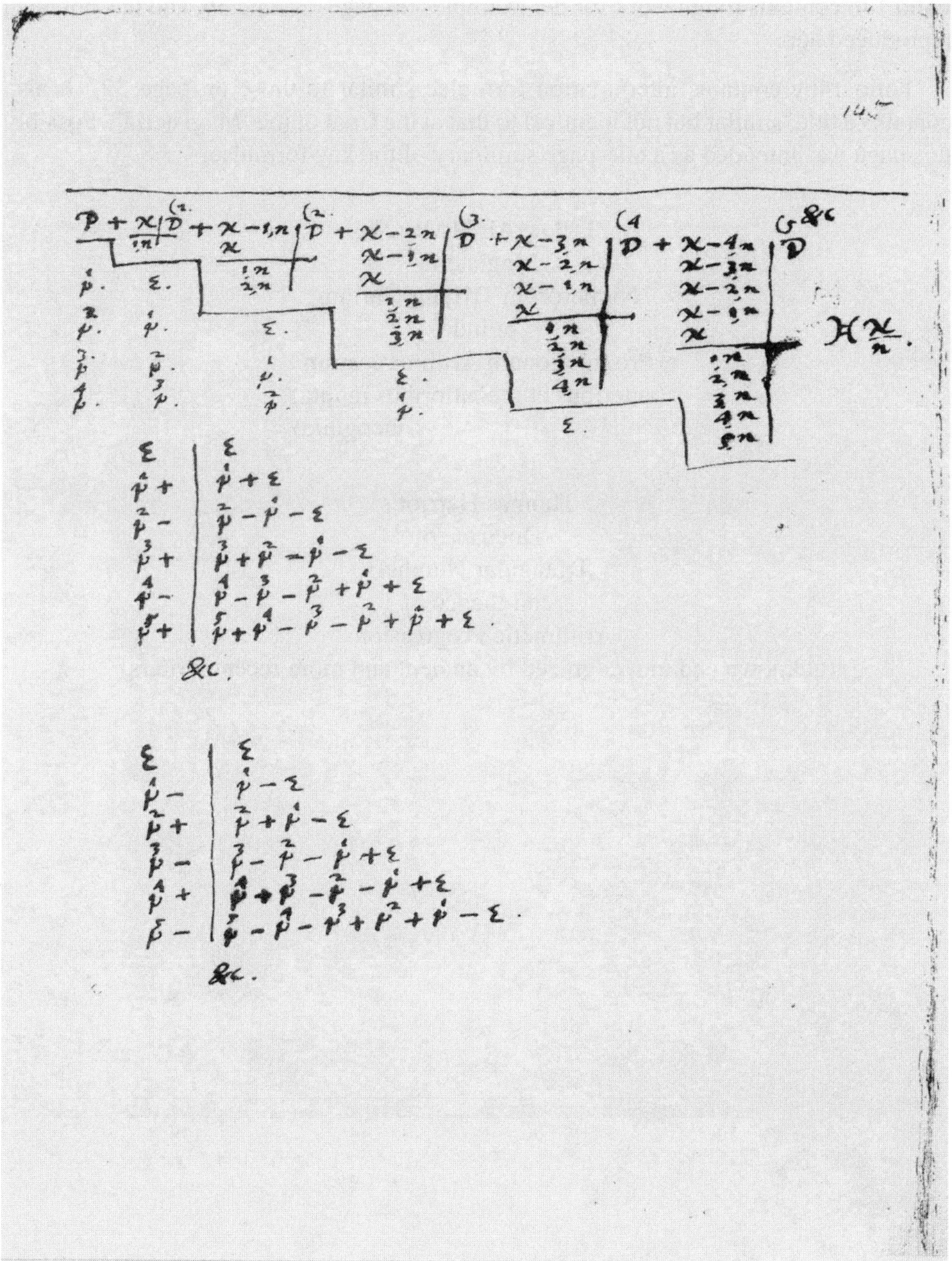

Endmatter [Folio 146v]

Folio 146 contains rough work for the examples on pages 35 and 36, and has not been reproduced here.

Folio 146v contains interpolation formulae similar to those on page 32. It also contains a title, similar but not identical to that at the front of the 'Magisteria'. Possibly this page was intended as a one-page summary of the key formulae.

THOMAE HARIOTI
Magisteria
Numerorum Tr[i]angularium
et inde
Progressionum Arithmeticarum
(veteribus et recentioribus ignota)
incognita)

Thomas Harriot's
Doctrine of
Triangular Numbers
and thence of
Arithmetic Progressions
(unknown and unrecognized by ancient and more recent writers)

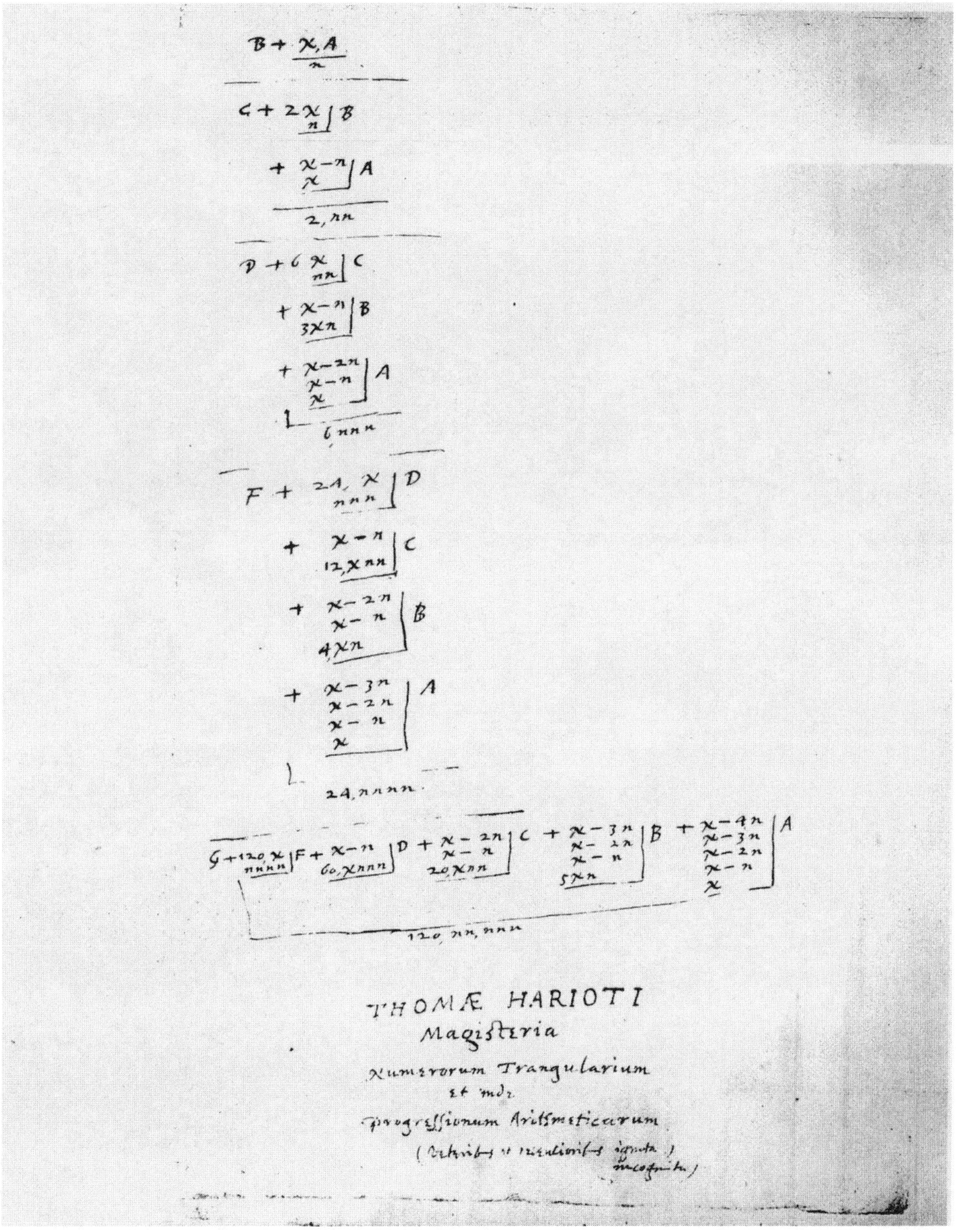

$B + \frac{x}{n} A$

$C + 2\frac{x}{n} B + \frac{(x-n)x}{2,nn} A$

$D + 6\frac{x}{nn} C + \frac{(x-n)3xn}{6nnn} B + \frac{(x-2n)(x-n)x}{6nnn} A$

$F + 24\frac{x}{nnn} D + (x-n)12xnn\, C + (x-2n)(x-n)4xn\, B + (x-3n)(x-2n)(x-n)x\, A \;/\; 24,nnnn$

$G + 120\frac{x}{nnnn} F + (x-n)60xnnn\, D + (x-2n)(x-n)20xnn\, C + (x-3n)(x-2n)(x-n)5xn\, B + (x-4n)(x-3n)(x-2n)(x-n)x\, A \;/\; 120,nn,nnn$

THOMÆ HARIOTI

Magisteria

Numerorum Triangularium

et inde

progressionum Arithmeticarum

Acknowledgements

The authors are grateful to the British Library, London, and to Lord Egremont of Petworth House, Sussex, for permission to use or reproduce pages from Harriot's manuscripts. We are also extremely grateful to librarians of the Bodleian Library, Oxford, (especially Colin Harris and his staff), Cambridge University Library, (especially Adam Perkins and Catriona Mackay), the British Library, Lambeth Palace Library, the West Sussex Record Office, Sheffield University Library, the Huntington Library, University of Delaware Library, and University of Redlands Library (especially Sandra Richey), all of whom have given us invaluable assistance. We also thank Peter Neumann of the University of Oxford for his careful reading of our work and his perceptive comments, and Janine Stilt of the University of Redlands for her help in preparing it for publication. Judith Tschann of the University of Redlands and Daniel Otero and Matthew Pitlyk of Xavier University of Ohio also offered helpful advice. Finally, we thank Irene Zimmermann for her skilful handling of an unusual and challenging piece of typesetting.